Science

YEAR 5

Sue Hunter and Jenny Macdonald

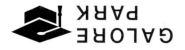

GALORE PARK

AN HACHETTE UK COMPANY

Acknowledgements p143 From Department for Transport and Driver and Vehicle Standards Agency, licensed under the Open Government Licence v3.0

Hachette UK's policy is to use papers that are natural, renewable and recyclable products and made from wood grown in sustainable forests. The logging and manufacturing processes are expected to conform to the environmental regulations of the country of origin.

Orders: please contact Bookpoint Ltd, 130 Milton Park, Abingdon, Oxon OX14 4SB. Telephone: +44 (0)1235 827827. Lines are open 9.00a.m.–5.00p.m., Monday to Saturday, with a 24-hour message answering service. Visit our website at www.galorepark.co.uk for details of other revision guides for Common Entrance, examination papers and Galore Park publications.

Published by Galore Park Publishing Ltd

An Hachette UK company

Carmelite House, 50 Victoria Embankment, London EC4Y 0DZ

www.galorepark.co.uk

Text copyright © Sue Hunter and Jenny Macdonald 2015

The right of Sue Hunter and Jenny Macdonald to be identified as the authors of this Work has been asserted by them in accordance with sections 77 and 78 of the Copyright, Designs and Patents Act 1988.

Impression number 10 9 8 7 6 5 4 3 2 1

2019 2018 2017 2016 2015

Typeset in India

Printed in Italy

Cover photo © mikelaptev - Fotolia

New Illustrations by Integra software services Pvt.Ltd

Illustrations p6, 9, 22, 2, 26, 78, 86, 87, 89, 93, 97, 98, 148, 157 by Ian Moores; p138 © Dorling Kindersley/Getty Images

A catalogue record for this title is available from the British Library.

ISBN: 9781471847516

Contents

About the authors

Sue Hunter has been a science teacher in a variety of schools for more years than she cares to remember. Her experiences have included teaching in a choir school and a local authority middle school, teaching GCSE and A level in the Netherlands and a short spell as a full-time mother of two. She was Head of Science at St Hugh's School in Oxfordshire until her recent retirement and is a member of the Common Entrance setting team. She has run a number of training courses for prep school teachers, including at Malvern College and for the Independent Association of Preparatory Schools (IAPS), and is currently IAPS Subject Leader for science and a member of the Independent Schools Inspectorate. She has also served for a number of years as a governor of local primary schools.

Jenny Macdonald has been a teacher for many years, teaching in both state and private schools. For the last 15 years she has been teaching science at St Hugh's School in Oxfordshire. She moved to Oxfordshire in the 1970s and has always enjoyed outdoor pursuits, having raised three children and countless chickens, sheep and dogs on the family smallholding. She is chairman of a local choral society, sings in a variety of local choirs, and would like to have more time to relax in the chairs that she enjoys re-upholstering.

Introduction

⇦ About this book

To the teacher

The scientist is not the person who gives the right answers, but the one who asks the right questions. Claude Lévi-Strauss

The study of science for young children is a voyage of discovery. It stimulates their curiosity and provides a vehicle for them to explore their world, to ask questions about things that they observe and to make sense of their observations. It does not exist in isolation but draws upon many other aspects of a well-rounded curriculum and should be practical, interesting and, above all, fun.

This book covers the requirements for the National Curriculum for Year 5. It also contains additional material as necessary to meet the specification for Year 5 in the ISEB Common Entrance syllabus and some extension material. It includes ideas for activities to develop practical skills, deepen understanding and provide stimulus for discussion and questioning.

Practical work is always popular and hands-on activities in this book are designed to be carried out by the pupils in pairs or small groups. Pupils should be encouraged to think about safety at all times when carrying out practical activities. However the responsibility for risk assessment lies with the teacher who should ideally try out each activity before presenting it to the class in order to identify any risks as appropriate to the particular group of children involved. The Association for Science Education (ASE) publication *Be safe!* (available via the ASE website: www.ase.org.uk/resources) is a useful source of information and advice about risk assessment in the primary phase.

Exercises have been set at intervals throughout the book. Where there is more than one exercise in a group, the first one is set at standard level followed by a more easily accessible exercise covering the same material and/or an extension exercise.

To the student

This book is to guide you in your study of science in Year 5. Science is a fascinating subject because it tells you so much about yourself and the world you live in. Science is all about asking questions and finding answers to them so use the information in this book as a starter but remember to look, listen and ask questions to take you further.

There are some special features in the book that are especially highlighted to help you in your work.

Notes on features in this book

Words printed in blue and bold are key words. All key words are defined in the Glossary at the end of the book.

Exercise

Exercises of varying lengths are provided to give you plenty of opportunities to practise what you have learned.

Activity

Sometimes it is useful to explore a topic in more detail by researching it. An activity is an opportunity to discover interesting things for yourself, and to practise recording and presenting what you find out. Some activities provide opportunities for you to do experiments. Others need some research from books or the internet, or maybe by talking to other people.

Did you know?

In these boxes you will learn interesting and often surprising facts about the natural world to inform your understanding of each topic.

Working Scientifically

Working Scientifically is an important part of learning science. When you see this mark you will be practising the really important skills that make good scientists. You will find out:

- why we carry out experiments
- what we mean by the word 'variable'
- what we mean by a fair test
- how to design experiments to answer your own questions
- how to measure variables
- how to record and display results clearly and accurately
- how scientific understanding is built up by the work of many scientists learning from each other, sometimes over hundreds of years.

Go further

The material in these boxes goes beyond the ISEB syllabus for 11+. You do not need to learn it for an 11+ exam but your teacher may decide that it is a good idea for you to learn something a bit extra to help you to understand a topic better or to extend your learning. All this material will be useful to you in your future studies …

1 Life cycles

⟨ Round and round

You have probably learnt about the life processes that are carried out by living things. Can you remember them all?

One of these life processes is reproduction. No living thing can live forever, although some can live for a very long time. It is therefore essential that all living things reproduce so that the species will continue living and not become extinct.

There are lots of different ways of reproducing. Some very small organisms are just one cell and reproduce by splitting themselves into two.

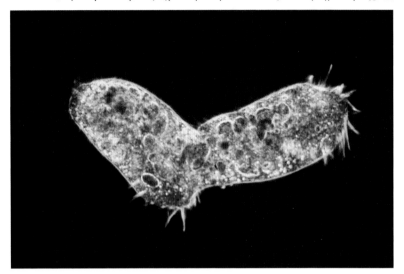

■ Single-celled organisms reproduce by splitting themselves in two

Most larger plants and animals reproduce by joining together two cells from different individuals. One of these special cells is usually called an egg or ovum and generally comes from the female parent. The male cell may be in a pollen grain in a plant or may be a sperm cell in an animal. These special reproduction cells are known as gametes. Each one contains half the information needed to make a new individual. The joining together of these two gametes is known as fertilisation and the resulting cell contains the whole set of information to make the new individual. This individual may grow to become an adult and then reproduce to create the next generation, and so on. This is what we call a life cycle. You have already learnt about how this process occurs in flowering plants. In this chapter you will learn about the life cycles of some animals.

Exercise 1.1

Use the words below to complete the following sentences. Each word may be used once, more than once or not at all.

cell   eat   extinct   fertilisation   gametes   sperm
one   ovum   reproduce   splitting   three   two

1 All living things need to _____ so that their species do not become _____.

2 Very small organisms that are made of just one _____ may reproduce by _____ the cell in two.

3 Most larger animals and plants need _____ special cells to reproduce.

4 The name given to special cells used for reproduction is _____.

5 The special cell that comes from the male parent is called a _____.

6 The special cell that comes from the female parent is called an _____ or _____.

7 The joining together of two _____ is called _____.

Activity – plant life cycles

1 Look at the diagram of a flower. Work with your partner or group to see if you can name the parts labelled A, B and C. Can you describe the function of each of these parts?

Clever plants

The plant life cycle you have learnt about involves the fertilisation of an egg cell by the male cell in a pollen grain. This is called **sexual reproduction**. Most animals also carry out sexual reproduction. Each new plant or animal will be a little bit different from its parents. You will learn more about why this is important later.

2 Take a sheet of paper or a fresh page in your book and draw a diagram showing the life cycle of a flowering plant. Label your diagram. Try to use the terms 'germination', 'pollination', 'fertilisation', 'seed formation' and 'seed dispersal' in your writing. Now compare your diagram with your partner's. Are there any differences? Discuss these to see if you can make a really good version by combining the best bits of both diagrams. Can you both explain clearly what each of these terms means?

3 The first two parts of this activity were about plants whose flowers are pollinated by insects or other animals. Do you know another method of pollination? How do the flowers of these plants differ from the insect pollinated ones?

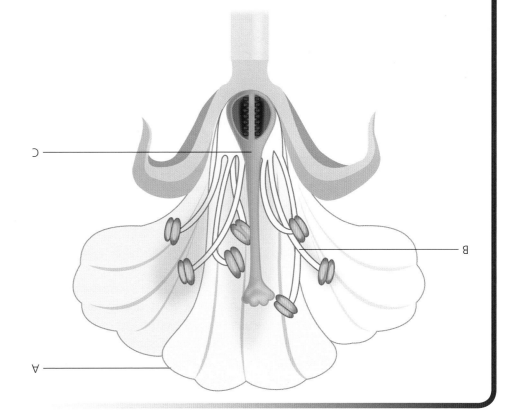

A

B

C

Many plants can also do something rather clever. They can produce new plants from their own stems, leaves or roots. The new little plant will be just like its parent in most ways. This is called **asexual reproduction**. Gardeners often make use of this fact to make new plants for their gardens. Maybe you have helped someone to take cuttings from a plant in the garden. If so, you have helped the plant to reproduce in this clever way.

■ Small pieces of stem, leaf or root, called cuttings, can be used to make new plants

Exercise 1.2

1 In a plant's life cycle give the word that is used to describe the following processes:

(a) when a seed begins to grow a root

(b) the transfer of pollen

(c) the joining together of a pollen cell and an egg cell in the ovary

(d) spreading seeds across a wide area.

2 What is the main difference between sexual reproduction and asexual reproduction?

← Animal life cycles

Very few animals are able to carry out asexual reproduction.

Some of the most interesting animal life cycles are those in which the animal changes its form completely when it becomes an adult. Can you think of any examples of this?

The Red Admiral butterfly

Moths and butterflies spend their early lives as caterpillars before changing into a beautiful winged adult form. Many other insects, such as flies and beetles, spend their early lives as grubs with no legs or wings. This huge change of form is called metamorphosis. Let's look at the life cycle of one of these animals in more detail.

The Red Admiral butterfly is often seen in our gardens during the summer, feeding on nectar from flowers. Most Red Admirals migrate to the UK from warmer places in Europe. They usually arrive in May or June, although some will have spent the winter in barns or sheds, sheltering from the cold. The adult

females lay their eggs on the leaves of nettle plants, and the young caterpillars hatch out after a week or so, depending on the weather. The caterpillars spend about three to four weeks feeding on the nettle leaves. When they are not feeding, they hide in little tents made by pulling the edges of a leaf together. As they grow bigger they make themselves bigger tents.

When they have grown big enough, they will pull several leaves together using silk they have spun, hide inside the tent they have made and fix themselves firmly to a leaf. Their skin hardens and turns a greyish brown colour. They stay like this for about two or three weeks. This stage of the insect's development is called a pupa. Inside the hard skin the animal gradually changes so that the body is no longer caterpillar shaped but has the three-part body of an insect. Three pairs of long, jointed legs appear on the thorax (the middle section of a three-part body). Two pairs of wings are formed, all crumpled up to fit inside the pupa's skin.

When all the changes have taken place, the skin splits open and the adult, or **imago**, pulls itself out. It pumps blood into the wings to stretch them out and spends some time sitting in the sun so that they dry and harden. The butterfly will then suddenly take wing to start a new life as a flying insect, feeding on nectar and finding a mate. Another set of eggs will then be laid and the life cycle begins again.

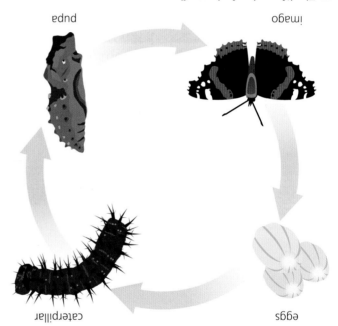

eggs

caterpillar

pupa

imago

■ The life cycle of a butterfly

Did you know?

It is often hard to tell the difference between a butterfly and a moth. Usually moths hold their wings stretched out flat and butterflies put theirs together above their backs when resting. Another way of telling them apart is to look at their antennae. If they are feathery, the animal is a moth. Butterflies usually have antennae like tiny drumsticks. Butterflies only fly in the daytime. Many moths are nocturnal.

Did you know?

Some species of moth, for example the spindle ermine moth, lay hundreds of eggs on one plant. When the caterpillars hatch out they spin a huge tent of silk over and around the bush they are on. They can then feed in safety, protected from predators.

Exercise 1.3a

1 Explain what is meant by the word 'metamorphosis'.

2 Name two animals that undergo metamorphosis in their life cycle.

3 What is the food plant for Red Admiral caterpillars?

4 How would you recognise a Red Admiral pupa?

5 Describe the changes that occur to the caterpillar while it is in the pupa.

6 What happens when all these changes have taken place?

Exercise 1.3b

Use the words below to complete the following sentences. Each word may be used once, more than once or not at all.

cabbage   imago   metamorphosis   nettle

pupa   sheds   tents

1 Some animals undergo a huge change in form, called —————— during their life cycles.

2 Red Admiral butterflies lay eggs on —————— plants.

3 The caterpillars protect themselves by making —————— out of leaves.

4 When the caterpillar is big enough it will fix itself to a leaf and turn into a ——————.

5 Inside the —————— the caterpillar changes form to become an adult butterfly, called an ——————.

Frogs

One group of vertebrates also undergoes metamorphosis in its life cycle. These are amphibians, most of which lay their eggs in water and have young that look very different to the adult. An example of an amphibian is the common frog, which is found in ponds all over the UK.

Adult common frogs are well adapted to life in water. They have webbed feet for swimming and eyes on the top of their heads so they can look out for prey or predators above the water, while keeping their bodies under the water. Their skin is specially adapted to allow them to absorb oxygen from the water so that they can spend long periods under water without breathing. In spite of all these adaptations, adult common frogs spend more time on land than in the water, living in damp places so that their delicate skin does not dry out, and feeding by catching flies with their long sticky tongues.

In spring, common frogs make their way to water, usually to the same pool where they began life. The males usually arrive first and begin croaking noisily to attract the females. When a female arrives the males compete to reach her. Sometimes a female may be mobbed by as many as ten males. She will lay her eggs, known as frog spawn, in the water

■ Common frogs return to ponds in the spring to lay their eggs in water

and the winning male will release a cloud of sperm at the same time. Most of the eggs will be fertilised and grow a protective layer of jelly around them. They can be found in huge clumps in some ponds. The jelly prevents the developing tadpoles from drying out and makes it a little harder for predators to eat them.

After about a week, a tiny comma-shaped tadpole forms and the jelly dissolves, releasing the tadpole into the water. At first the tadpole feeds on tiny water plants called algae, scraping them off the surface of plants with a rasping mouth which is on the

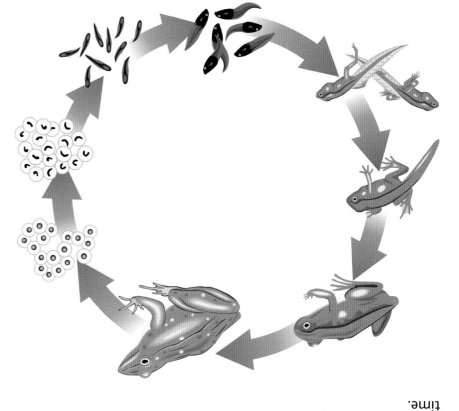

The life cycle of a frog

12–14 weeks, a tiny frog emerges from the water for the first time.

About seven weeks after hatching, the back limbs begin to grow. These are tiny and useless at first, but they gradually grow bigger and stronger. Soon the front legs also begin to grow, the eyes become bigger and the tail starts to shrink, and the young frog develops lungs to breathe in air. Eventually, at the age of about

water fleas and other invertebrates.

At around this time the tadpole's diet also changes and it begins to feed on tiny water fleas and other invertebrates.

After a few weeks, the tadpoles begin to grow legs

underside of its head. These tiny tadpoles have feathery gills on the outside of their bodies but after a week or two these are replaced by internal gills.

Animal life cycles

Did you know?

The axolotl is a type of amphibian found in Mexico. It starts its life cycle like other amphibians by laying eggs, which hatch into tadpoles. The tadpoles of the axolotl grow front and back legs but they never complete the metamorphosis, and remain in the water breathing through the gills they keep. There may be fewer than 1000 of these curious animals left in the wild although, as they are popular pets, many exist in captivity.

Exercise 1.4a

1 To which group of vertebrates does the frog belong?

2 Where do frogs lay their eggs?

3 What name is given to frogs' eggs?

4 What differences are there between a tadpole and an adult frog?

Exercise 1.4b: extension

1 A pond will often contain frog spawn from many different frogs.

(a) Suggest one reason why it might be good for the baby frogs if thousands of eggs are laid in the same pond.

(b) Suggest a reason why the presence of so many tadpoles in the same pond might be a disadvantage.

2 In the rainforest, tree frogs spend their lives in the branches of trees and seldom come to the ground. Where do you think they might lay their eggs? (You could do some research to find out if you do not know the answer.)

Birds

Birds also lay but their eggs have a hard, chalky shell around them to protect the growing baby. Most birds will construct a nest to lay their eggs in and then the parent birds will sit on the eggs to keep them warm until they hatch. Baby birds look like small featherless adults. There is no metamorphosis in their life cycles. Let's learn a bit more about the life cycle of one particular bird.

You will probably recognise this bird easily. The robin is one of our best-known birds and is seen and heard in gardens and parks all the year round. Robins often build their nests in undergrowth but will happily nest in bird boxes. They are also known for building nests in all sorts of strange places such as post boxes, coat pockets or hanging baskets.

Robins are very territorial. They will often be seen perched on a tall tree singing loudly to tell all the other robins to keep out of their patch. Around March time, a male and a female will come together in one territory to breed. The female builds her cup-shaped nest out of dead leaves and moss. The pair will then mate and she will lay her small blue eggs, one each morning for about four to six days. She will then sit on them for about 13 days. The male brings her food during this time.

When the eggs hatch, the little birds have no feathers and their eyes are closed. It takes more than ten days for all the feathers to grow but their eyes begin to open after about five days. Both parents are now very busy catching insects to feed the chicks.

When all their feathers have grown, the baby birds will fledge. This means that they leave the nest. They are not very good at flying yet and so this is a very dangerous time for them. The parents will still feed them for about three weeks until they are strong enough to look after themselves. You may then see them searching for food in the undergrowth or on a bird table. Youngsters are easy to spot because they are a speckled brown colour. Their bright red breast feathers take several weeks to develop. The parent birds will often build a new nest and raise another brood of babies. They may have up to four broods in a good year.

When winter comes the young robins will need to find their own territories to feed in. The following year they will find mates and have babies of their own.

Did you know?

There is a bird in North America that is also known as a robin. It is not actually a robin but a type of thrush. It is thought that the early settlers called it a robin because its red breast reminded them of the friendly birds they knew from home in Britain.

Exercise 1.5a

1 At what time of year will a pair of robins first come together?

2 Describe a robin's nest.

3 How many eggs will the female robin lay in her nest?

4 How long does it take for the robin's eggs to hatch?

5 What do newly hatched robin chicks look like?

6 How long does it take for the baby robins to grow their feathers?

7 What type of food will the parent birds collect for the babies?

8 What does the word 'fledge' mean?

9 Why are the first few days after fledging so dangerous for the baby robins?

10 How can a young robin be distinguished from an adult?

Exercise 1.5b

Find words in the text about birds to help you to complete the following passage:

A robin's nest is made from _____ and _____.
The female robin will lay about _____ eggs. The eggs are coloured _____ and have _____ shells. It takes about _____ days for the eggs to hatch. The baby robins have no _____ and their eyes are _____. The parents catch _____ to feed the babies. When the baby birds' feathers have grown they will leave the _____ but they are not very good at _____ yet. The parents will still feed them for about _____ weeks. Young robins are speckled _____ in colour and do not get their _____ breast feathers for a few more weeks.

Exercise 1.5c: extension

1 What do we mean by the word 'territory'?

2 Robins are territorial, which means that they defend their territory and may be quite aggressive to other robins. Suggest what would make a good territory for a robin. Think about what the robin might need, especially in the winter and during the breeding season.

⇨ People in science: Jane Goodall and David Attenborough

Have you ever wondered how we know so much about the animals and plants that share our world? There are two groups of people who have very important roles in helping us to find out about the world around us: the scientists who study the natural world and those people who communicate their findings to the world in a way that everyone can understand.

Jane Goodall and the chimpanzees of Gombe

Many scientists work in laboratories making observations and doing experiments. This works quite well for small animals and plants and we have learnt a lot about the lives of these organisms in this way. However we also need to know about how animals behave in their natural habitats and the only way to study this is to share their habitat over a long period of time. It takes a very special kind of person to do this. One such person is the animal behaviourist Jane Goodall, who has spent most of her life following a group of chimpanzees in the Gombe Stream area of Tanzania in West Africa.

Jane first went to Africa to study the chimps in 1960, when she was 26 years old. She had always been interested in animals and this was a big adventure for her. At first the chimpanzees ran away when they saw her but she continued to watch them patiently  from a distance. She made notes of everything she saw them do and took photographs as well. Gradually the chimpanzees began to understand that she was not a threat to them and they started to accept her. This meant that she could watch them more closely and get to know them as individuals.

When Jane first started studying the chimpanzees, people thought that humans were the only animals that used tools. One day Jane saw a big old male chimpanzee, whom she had called David Greybeard, pick up a twig and strip the leaves off it. He then used the twig as a tool to fish termites out of their nest. This was the first time an animal had been seen using a tool. Another important observation that she made was that chimpanzees often eat other animals such as monkeys and bush pigs. Before this people thought that chimpanzees were almost completely herbivorous.

■ Chimpanzees were the first animals to be seen using tools

As well as observing the chimpanzees' feeding behaviour, Jane watched the ways in which they behaved towards each other. She was shocked to find that groups of chimpanzees will engage in wars against each other. However, she also saw how these animals form strong family bonds and care for each other in a very similar way to human families. She even watched young male chimpanzees caring for two babies whose mothers had been killed.

To find out all these interesting new facts about chimpanzees took many years of patient observation. Jane watched many baby chimpanzees grow up, learning new skills from their parents and other members of the group and eventually mating and having babies of their own. She could see how well they were adapted to their life in the forest and how they used their intelligence to solve problems to help them survive. However, she realised that these wonderful animals are threatened by the destruction of their habitat. She now travels the world speaking to groups of people about what can be done to preserve the forests that are home to the chimpanzees and many other animals and plants. Thanks to her dedicated work, we now know so much more about these animals and their habitat and have the opportunity to make sure that their future is secure.

Did you know?

At the beginning of the twentieth century there were between 1 million and 2 million chimpanzees in the world. Now there are only around 300 000.

Activity – conservation

Read the following information about the threat to chimpanzee habitats:

The survival of the chimpanzees at Gombe and in other parts of Africa is at risk for many reasons. Local people need farmland to grow crops to feed their families and to earn money so they cut down some of the forest. They also need firewood and many of them rely on hunting wild animals, including chimpanzees, for food (bushmeat). The forest is their home and provides almost everything they need.

All over the world, people use a lot of timber and big companies cut down trees in the forests to provide it. This process is called logging. In many areas of chimpanzee habitat there are also valuable minerals in the rocks and so companies want to mine these to provide the materials we need to supply the industries of the world. Logging and mining provide jobs and income for lots of people but cause great damage to the forest.

Conservationists and governments need to understand all these different factors and try to find ways to protect the forest whilst allowing for the needs of the local people for whom the forest is their home.

Imagine that a large international company wants to dig a mine in an area of chimpanzee habitat.

Your class will be divided into three groups. One group should represent the mining company, one group should be the local tribe of people who live in the forest and a third group should be the conservationists who want to protect the chimpanzees. Your teacher could be the official from the government who has to decide whether to let the mining begin.

In your group do some research about your group's point of view and prepare a short presentation. Remember to think about what is important to each of the other two groups and try to make your presentation as persuasive as possible.

Now present your ideas to the class and to your teacher and then have a debate to discuss the ideas and ask questions. At the end your teacher can decide which group has been most convincing and whether the mining should happen or not.

Activity – chimpanzee research project

This chapter has given you some facts about chimpanzees but there is a lot more that you could find out.

Start by writing down about five or six questions that you want to find out about. One of these should be: How are chimpanzees adapted for life in the forest?

Next, discuss with your partner or group what you think would make a good piece of research work. Decide on about three or four assessment criteria. For example, you might choose: Does the research answer the questions? Is the information clearly presented?

Use books and the internet to gather information to answer your questions. Make sure that your information is well-organised and accurate by checking the facts in more than one place.

Prepare a fact sheet or poster to display your information. Make sure that your initial questions are clearly shown and that your information answers them. You could draw or print out pictures, maps and diagrams to illustrate your work. Remember that all your writing must be in your own words and remember to include a list (bibliography) of the books and websites where you found your information.

When you have finished, work with a partner or small group to assess each other's work, using the criteria you agreed at the beginning. Remember to be positive. Start with what you really like about their work and then make a helpful suggestion about how it might be improved.

David Attenborough – a life in broadcasting

Scientists who carry out research projects report their findings in many different ways. Often they write articles for scientific publications and these may be very complicated and hard to understand. We need people who can take this information  and present it to us clearly and simply so that we can learn more about our world. One of these people is Sir David Attenborough.

As a child David was always interested in nature and was particularly fascinated by fossils. He studied geology and zoology at Cambridge University. David is a naturalist, which means he is an expert on wildlife.

In 1952, he joined the BBC. At that time almost all broadcasting was by radio. Television was a very new technology. David started work as a producer but soon started to appear in front of the camera as a presenter of programmes about wildlife. He later became the controller of BBC2 for a while, but he continued to make programmes about wildlife all round the world. He always wanted to show his audience the most fascinating things about animals and plants and became a pioneer in the field of natural history broadcasting.

When David started working on television broadcasts it was not really possible to make films outside the studio. Cameras were huge and all programmes were broadcast live because the recording technology was not good enough to record and edit pieces of film. All television was in black and white. However, David has always been keen to make the most of new technology as it becomes available. As soon as it became possible to do so, he started to film animals in the wild and he made some of the very first colour television programmes. His broadcasts have often included film made with new equipment and this has allowed him to show us features of animals and plants in increasing detail over the years. Often he and his film team have been the first to film some aspects of animal behaviour.

■ Sir David Attenborough always makes the most of current technology in his films

David has been making wildlife films for over 60 years. He has shown us animals and plants from many different habitats all round the world, sharing his personal enthusiasm and allowing us to sit at home and see the wonders of the natural world almost as if we were actually there with him. To do this he has climbed huge trees, entered pitch-dark caves full of bats and cockroaches and dived under the oceans. He has worked in some of the hottest and coldest places on Earth and has spent long periods of time patiently watching and waiting to get the best possible shots. He has won many awards for his work.

Like most people who study organisms in their habitats, David has become concerned about the ways in which human activity threatens the natural world and often includes this in his broadcasts. His work is seen by people in many countries all round the world. His skill in communicating and the skills of the cameramen and women and all the other people who work with him have been very important in teaching people everywhere about the world around us.

Did you know?

When David Attenborough was a boy he collected fossils and rocks and other interesting natural specimens. He found out more about them and created his own mini natural history museum. Maybe you could do the same.

Activity – make a wildlife broadcast

Watch a short part of one of your favourite David Attenborough programmes. Discuss what makes a good communicator.

Prepare a short talk about a natural history topic. It could be about a particular animal or habitat or maybe a conservation message.

Present your talk to the class. If you have a video camera or other recording equipment, you could ask someone to record your talk so that you can use your IT skills to turn it into a television or radio broadcast.

2 More about life cycles

■ Human babies look like a smaller version of their parents

Have you ever seen a lamb being born or a chick hatching from an egg? Maybe your dog has had puppies. You may have watched a film of baby animals being born. Many baby animals look similar to their parents. They may be blind and hairless or have fluff instead of feathers but, in general, they are the same shape, just rather smaller. They will grow without any major change, or metamorphosis, into adults.

⇨ The human cycle

When a human baby is born, people often look to see if it looks most like its mother or its father. In fact, a baby will be like its mother in some ways and like its father in other ways. To understand why, we need to think about how babies are formed and what happens as a baby grows up.

A newborn baby is pretty helpless. It doesn't seem to do much other than eat and sleep, but it is doing a lot of learning. It needs to learn how to feed, how to recognise its parents and how to tell them when it is hungry, tired or uncomfortable. It looks around all the time as its brain begins to make sense of the images it sees and the sounds it hears. It is also growing fast. At first, its food is only milk, but after a while it will begin to eat solid foods too. It will learn to smile, try to communicate and begin to move around, to play, to feed itself and to walk. All this time, it needs adult care.

Toddlers may begin to go to places away from home where they can learn to play with other children and, after a few years, they will probably go to school. All this time, they are growing and learning to do new things.

Adolescence

As girls and boys grow up, they begin to notice changes happening in their bodies. These changes begin at different ages in different people but sooner or later they happen to everyone. The stage in our lives when these changes take place is called adolescence. Adolescence is the time when our bodies begin to get ready for adult life.

Some of the changes are particular to boys and some to girls. Boys become stronger and their voices 'break' (become deeper). They begin to grow hair on their faces and their penises become larger.

Girls may notice that their breasts begin to get larger. Breasts are important because they are the place where milk is made for human babies. Young girls do not need large breasts but the breasts begin to grow in adolescence so that when the girl is old enough to have children of her own she will be able to provide the nutritious milk that a baby needs for healthy growth. Other changes also happen inside girls' bodies at this time.

Inside a woman's body there are two ovaries where eggs are made. These are joined to a place called the uterus or womb by two tubes, called fallopian tubes or oviducts. The uterus is the place where a baby grows until it is big enough to be born. When a girl goes through adolescence, the uterus begins to develop a thick lining, full of blood vessels. If a baby were inside the uterus, these would provide the oxygen and nutrients it would need.

It is really important that the uterus is perfect for a baby to grow in, so the uterus refreshes itself monthly. Each month one of the ovaries releases an egg into the tube connecting it to the uterus. If the egg is not able to develop into a baby, it passes out through

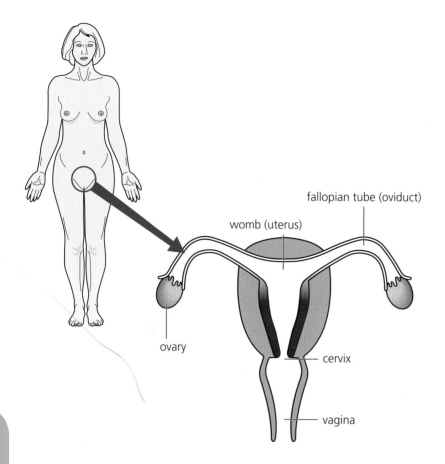

fallopian tube (oviduct)

womb (uterus)

ovary

cervix

vagina

the opening called the vagina and the thick lining of the uterus begins to break away. The blood that was in the lining passes out of the body through the vagina. This monthly bleeding is called a period and usually lasts for a few days. During this time a girl will need to use sanitary towels or tampons to absorb the blood. The lining of the uterus is then replaced with a new one ready for the next egg. This monthly cycle of changes to the uterus is known as the menstrual cycle.

Some changes happen to both boys and girls. Some people find the skin on their faces becomes greasier and they may get spots. Most people find that the smell of their bodies changes, especially when they get a bit hot and sweaty. Hair grows thicker in certain places on their bodies. Many young people become rather moody during adolescence and they begin to care a lot more about what other people think of them. It can be a very bewildering time and it is important to understand what is happening so that everyone can be supportive of each other.

Exercise 2.1a

1 Describe some of the things that a human baby needs to learn in the first few months of its life.

2 What name is given to the part of a human's life when they change from being a child to being an adult?

3 Describe some of the changes that happen to boys' bodies during the time you have named in question 2.

4 Girls find that their breasts grow bigger when they grow up. What is the function of a woman's breasts?

5 What is happening when a girl has a period?

6 Describe some other changes that may be experienced by both boys and girls as they go through this time of change.

Exercise 2.1b

Use words below to help you to complete the sentences. Each word may be used once, more than once or not at all.

care	deeper	hair	higher
milk	month	moody	adolescence
uterus	sleep	stronger	year

1 Newborn babies need a lot of _____ and cannot _____ for themselves.

2 The part of our lives when we change from being a child to an adult is called _____.

3 When boys grow up, their voices become _____ and their bodies become _____.

4 When girls grow up, their breasts grow larger so that they can produce _____ to feed their babies.

5 A period is when the lining of the _____ is renewed. This happens about once a _____.

6 During adolescence boys and girls may become _____ and they will grow more _____ in certain places on their bodies.

Making a baby

As with most animals and plants, making a human baby takes two parents: a mother and a father. A baby can only start to grow if one of the eggs from the mother's ovaries joins together with a sperm cell from the father. The egg and the sperm cell each contain half a set of information for making a new person. When they join together a full set of information is created, half of which comes from the baby's mother and half from the father. This is why each baby is a bit like both its parents.

Sperm cells are made in two special places in a man's body, called testes. When a man and a woman have decided that they want to have a baby, they get very close together and the man pushes his penis into the woman's vagina. The sperm cells rush out of the testes, though the penis and into the woman's vagina.

Sperm cells have tiny tails and are able to swim. They swim up through the uterus in search of the egg. When a sperm finds the egg it burrows in and joins with the egg. This is called fertilisation.

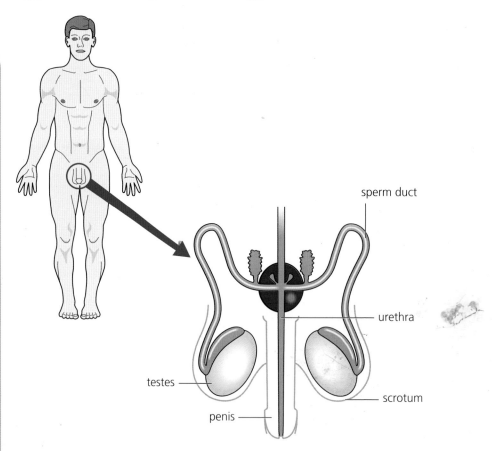

sperm duct

urethra

testes

scrotum

penis

The two special cells have now become one cell. This will begin to divide into two, then four, then eight and so on. Eventually a little ball of cells will reach the uterus and bury itself in the thick lining. This little ball of cells develops over nine months into a baby.

Did you know?

Sperm cells are the smallest human cells. They are about 0.05 mm long, including the tail. Sperm cells are too small to see with the naked eye. Egg cells are much bigger. They measure about 0.2 mm and are therefore just big enough to see without a microscope. Can you suggest any reasons for this difference in size?

The growing baby

When the ball of cells first reaches the uterus it is less than 1 mm long. The different parts of the baby soon begin to develop and by about 12 weeks it is a tiny little human, just 5–6 cm long and weighing 15 g. Although it is so tiny, it is already possible to tell if it is a boy or a girl, it can make a fist and put its thumb in its mouth. By 20 weeks the baby is moving around and can hear sounds. It has grown to about 300 g and is about 25 cm long from head to heel.

The baby will now grow very quickly indeed. All the different parts of its body will develop, it will begin to respond to sounds and by about 24 to 26 weeks the baby would have a good chance of surviving outside the uterus. By now it is about 26 cm long and weighs about 750 g. Most babies continue to grow inside their mother's body for about another three months and will be born when they are, on average, about 50 cm long and weigh 3.5 kg.

When a baby is developing in the uterus, it is known as a fetus. As the fetus grows, it needs plenty of oxygen and nutrients. It cannot breathe and eat for itself so the mother has to do these things for it. A structure called the placenta is created to join the growing fetus to the thick lining of the uterus. The fetus is attached to the placenta by the umbilical cord. Oxygen and nutrients from the mother's blood can pass into the baby's blood through the placenta. Carbon dioxide and other waste materials from the fetus are passed into the mother's blood so that the mother's body can get rid of them. The mother's blood and the baby's blood do not mix; materials are passed through the walls of the blood vessels from one to the other.

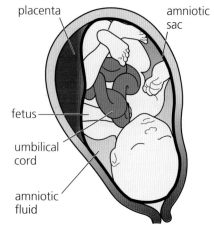

placenta

amniotic sac

fetus

umbilical cord

amniotic fluid

■ The baby in the uterus is provided with oxygen and nutrients through the placenta and umbilical cord

To protect the fetus from injury while it is in the uterus, it is surrounded by a bag of watery fluid called the amniotic sac. The growing fetus floats in this liquid. It also swallows it and takes it in and out of its lungs, helping to develop the muscles it will need to eat and breathe when it is born. This liquid also helps to keep the fetus at a constant temperature.

Did you know?

In the early stages of development it is very difficult to tell the difference between the fetuses of humans and all other mammals. Even more surprisingly, early stage mammal fetuses are also almost identical to those of birds, reptiles, amphibians and fish.

Birth

A human baby takes about nine months to develop in the uterus before it is ready to be born. When the time is near, most babies will turn upside down so that they are born head first. The mother will begin to feel the muscles around the uterus contract and relax as they begin to push the baby out of the uterus. The narrow passage through the cervix and vagina will have become quite stretchy so there is room for the baby to be squeezed out into the world. When the baby is born, a nurse or doctor will cut the umbilical cord. This does not hurt the baby because there are no

nerves in the cord. The remains of the cord become the baby's belly button. The baby will be encouraged to take its first breaths and soon will be happily feeding on milk from its mother's breast. A new person has been born.

Activity – us as babies

Although the average length and mass for babies is given above, all babies are different and some will be bigger and some smaller than the average.

Make up a questionnaire to find out the mass your friends were when they were born. Show the results in a bar chart. You might like to find out some other facts about them, such as the colour of their eyes when they were born or whether they had any hair. Maybe you could also collect some photographs and make a wall display.

Exercise 2.2a

1 What name is given to the structures in a woman's body that produce eggs?

2 Where in a man's body are sperm produced?

3 What is meant by the term 'fertilisation'?

4 Where in the mother's body does the baby grow?

5 How long does it take for a human baby to develop in its mother's body before it is born?

Exercise 2.2b

Complete the sentences below by selecting the correct word from each set of brackets.

1 In a woman's body, eggs are made in the (ovaries/uterus).

2 Sperm are made in a man's (ovaries/testes).

3 To start to make a baby, an egg cell must join with a sperm cell. This is called (fertilisation/reproduction).

4 The baby develops in the (uterus/vagina) in the mother's body.

5 A baby spends about (six/nine/twelve) months inside its mother's body before it is born.

⇨ Other animals

Humans belong to the group of animals called mammals. Other members of this group have a life cycle that is very similar to ours. Two parents mate and the baby animal grows inside the mother's body in the uterus. The baby animal is born in a similar way and is fed on milk from its mother.

One thing that differs in these animals is the length of time that the baby spends in the uterus. This time is called the gestation period. In humans, the gestation period is about nine months, but a baby elephant spends nearly two years in the uterus before it is born. Small mammals have very short gestation periods. For instance, baby hamsters take only about 16 days to develop enough to be born.

Human babies cannot do very much for themselves when they are born. If you have ever seen newborn kittens or puppies, you will know that they are even more helpless at birth than human babies and their eyes do not open at all until some time after they are born. There are also babies born to some animals that are able to stand and even run within hours of being born to help them escape predators.

■ Kittens cannot see or hear after they are born

Activity – human life cycle

The human life cycle has many stages and these are all given names: infant, child, adolescent (teenager), adult, elderly person.

Discuss these different stages with your partner or group and then draw a timeline for a human.

People often become less active and frailer as they get older. Suggest how this might make life more difficult for them. In many cultures old people are considered to be very wise and are greatly respected for this. Why might an older person be considered in this way?

Activity – animal babies

Choose one of the animals whose life cycle you learnt about in Chapter 1. Read the information about this animal again to remind yourself of the details. You are going to compare this animal's life cycle with our own.

Now draw a table with two wide columns. Write the heading 'Similarities' at the top of the first column and 'Differences' at the top of the second.

In the first column write down anything that is the same in both life cycles, (for example, both animals have two parents) and in the second write down any differences, (for example, frogs lay eggs in water but humans give birth to babies). Compare your lists with other members of the class. Are there any features that occur in all life cycles?

Activity – mammal life cycle

Humans are mammals and we share our lives with many other animals that belong to the same group. Most of our pets such as dogs, cats, hamsters, horses, and farm animals such as cows and sheep, are mammals. the life cycle of all mammals is similar to our own.

Choose one mammal, a pet or farm animal, or a wild mammal, and find out about their life cycle using books and the internet. You may need to use the term 'reproduction' in your search. Make a leaflet or poster showing your findings. Remember to give the details of any book or website you used and make sure that your work is all in your own words.

3 Adaptation and habitats

⇨ What is a habitat?

A habitat is a place where an animal or plant lives. The group of plants and animals that live there is called a community. In order for a community to live and grow well, the habitat must provide three important things:

- It must provide a place to feed.
- It must provide a place of shelter.
- It must be a place where the animal or plant can reproduce.

Anywhere can be a habitat; it might be under a log in your garden, it might be a park, a pond, a wood, a cemetery, the land beside a motorway or railway, or even a rubbish dump! Provided the animal can find food, shelter and be able to reproduce there, then it is a habitat.

Small habitats are often part of larger habitats. The habitat under a rock might be beside a pond, and the pond might be in a woodland.

How do habitats differ?

Some habitats, such as tropical forests, provide homes and feeding places for many different kinds of plants and animals. Other habitats, for example, the Arctic, provide homes and feeding places for fewer kinds of plants and animals.

■ A tropical rainforest

Some plants and animals can live very successfully in many different habitats. Plants and animals grow well in habitats where they are well adapted. For example, rabbits often live in large numbers on farmland, and yet they seem equally at home living in cemeteries, on open hillsides or even by the roadside. These are good habitats for rabbits because rabbits are adapted to eating grass and digging burrows to shelter and raise their young. Foxes might also live in all these different places because they are adapted to catching rabbits to eat.

Sometimes animals are found in very different habitats but behave differently in each one. For example, we often find gulls at the coast, flying over the sea and catching fish. However, we also find gulls inland a long way from the coast, feeding on rubbish that they find in rubbish tips, or following behind tractors and catching worms turned up by the plough. They have found ways of living successfully in different habitats, and we say they have changed their behaviour to find food in these different habitats.

■ Gulls pursuing a tractor in a half ploughed field

Some animals move from place to place between habitats. We call this migration. Sometimes animals migrate because the food supply in one place has run out so they travel to somewhere where more food is available. In Africa, for example, huge herds of wildebeest and zebras can be found travelling in search of fresh grass. Birds such as swallows will spend the summer in one place where they can find plenty of food to raise their young but then fly thousands of miles to Africa to escape the cold of winter.

■ Wildebeest in Kenya migrate long distances in search of food

For a plant to grow successfully in its habitat it needs space to grow. It must be able to take the moisture and minerals it needs from the ground. It also needs to be able to take enough light energy from sunlight to make its own food. Some plants can only grow in particular places. For example, gorse bushes grow best in sandy soils but flag irises do best in marshy areas beside ponds; ferns grow well in shade but sunflowers do better in bright sunny conditions. Other plants can grow in many kinds of habitats. For example, nettles will grow in good soil or in rubble or waste ground as long as they have enough light. Grasses grow almost everywhere and can survive well even when they are grazed by animals.

Exercise 3.1a

1 Explain what is meant by the word 'habitat'.

2 What do animals need in a habitat if they are to survive?

3 What do plants need in a habitat if they are to survive?

4 What is the name given to the group of plants and animals living in a habitat?

5 (a) Name three habitats where you might find gulls.

  (b) What sort of food might gulls be feeding on in each of these habitats?

6 What word is used to describe the way in which some plants and animals are able to change in order to survive in their habitats?

7 What is meant by the term 'migrate'?

Exercise 3.1b

Use the words below to fill in the gaps in the sentences. Each word may be used once, more than once or not at all.

adapted anywhere community feed migrate reproduce shelter

1 In order for a habitat to be successful, the plant or animal must be able to _____ , _____ and _____ .

2 The plants and animals in a habitat are called a _____ .

3 Almost _____ can be a habitat.

4 Animals and plants live and grow well in habitats that they are well _____ to.

5 Some animals _____ between habitats to find food or to breed.

Looking closely at habitats

If you are quiet and look and listen carefully when you visit a habitat, you will discover that there are many different animals and plants for you to study. On the following pages there are pictures of two habitats you may have seen. You can read about just a few of the living organisms that can be found in them.

The first habitat is a cemetery, which can be found in the town, city or countryside. A cemetery is usually a peaceful place, and provides animals and birds with a quiet area in which to feed, find shelter and reproduce.

The second habitat is a rock pool. You may have visited the seaside and spent some time looking in rock pools on the beach. A rock pool is a very exciting habitat. Here you will find some plants and animals that cannot survive out of water – they are safe in the pool when the tide has gone out. When the tide comes in again some of the animals will escape, others will arrive, and some will stay. So the community will be constantly changing.

Exercise 3.2a

Look at the picture of a cemetery habitat and use the information to help you to answer these questions.

1 Give the proper name for the homes of the following animals: badger, red fox, squirrel, rabbit.

2 Explain in your own words the meaning of the word 'hibernation'.

3 Name the animals in the cemetery picture that hibernate.

4 Name all the animals in the picture that have earthworms in their diet.

5 Describe how the red fox is adapted to help it hunt successfully.

6 Suggest why it is important for rabbits to have good eyesight and hearing.

7 What is the meaning of the word 'nocturnal'?

8 Which of the animals in the picture are nocturnal?

9 The owl has very good eyesight but the badger has poor eyesight. Suggest an explanation for this observation.

Bats
(Mammal)
Nocturnal and sleep in roofs, caves or holes in trees during the day.
Feed on insects.
Make high-pitched sounds and listen for the echo to help them to fly around in the dark.
Hibernate in winter.

Oak tree
(Deciduous tree)
Makes its own food using energy from the Sun.
Provides food and shelter for many insects, birds and small mammals.
Squirrels like to eat the acorns.

Red fox
(Mammal)
Feeds on many things, including rabbits, wood mice, earthworms and blackberries.
Most active at night.
Has excellent sight, hearing and sense of smell.
Lives in an underground den called an earth.

Wood mouse
(Mammal)
Nocturnal.
Feeds on seeds, fruits, earthworms, fungi and acorns.
Has large eyes and a good sense of smell.
Nests in underground holes.

Grass
Makes its own food using energy from the Sun.
Food for many animals including rabbits and insects.

Barn owl
(Bird)
Feeds on mice and shrews.
Hunts mostly at night.
Has large eyes and good hearing and can fly silently.
Nests in holes in buildings or old trees.

Badger
(Mammal)
Feeds on earthworms, fruit, roots and invertebrates.
Nocturnal.
Has good hearing and sense of smell but not very good eyesight.
Makes an underground den called a sett.
Has strong claws for digging.
Is not very active in winter but does not hibernate properly.

Rabbits
(Mammal)
Feed on grasses and other plants.
Have good eyesight and hearing.
Live underground in holes called warrens.

Fungi
Are not plants because they cannot make their own food.
Decompse dead plants and animals and return minerals to the soil.
Provide food for small animals.

Yew tree
(Evergreen tree)
Makes its own food using energy from the Sun.
Fruits contain seeds that are poisonous to people.
Birds and animals can eat the fruit safely.

Squirrel
(Mammal)
Feeds on seeds, nuts and buds and stores food for the winter.
Makes a nest called a drey in trees.
Sleeps a lot in winter but does not hibernate properly.

Blackberry bush
Makes its own food using energy from the Sun.
Prickly bush with sweet juicy fruits.
Leaves and berries are eaten by many birds, mammals and invertebrates.
Safe place for nesting birds.

■ Cemetery habitat

Exercise 3.2b

Look at the picture of a cemetery habitat and use the information to help you to answer these questions.

1 Match the following animals to their homes.

Animals	Homes
badger	warren
rabbit	drey
red fox	earth
squirrel	sett

2 What is the meaning of the word 'hibernate'?

3 Name one animal in the picture that hibernates.

4 What do foxes eat?

5 Why do rabbits need good eyesight and hearing?

6 What word do we use to describe animals that come out at night?

7 Name two animals in the picture that are seen at night.

Exercise 3.2c: extension

1 Find out how owl feathers are different from other bird's feathers, so that they are able to fly silently at night.

2 (a) Bats are able to fly around in the dark without bumping into anything. Find out how they are able to do this.

(b) What is the name given to this process?

(c) Suggest why submarines use a similar method for navigating deep in the ocean.

Did you know?

Hermit crabs (see the picture of the rock pool habitat) are often found with sea anemones on their shells. The anemone helps to camouflage and protect the crab and will eat leftover scraps of the crab's food. When the crab moves to a new shell, it may move the anemone from one shell to the other.

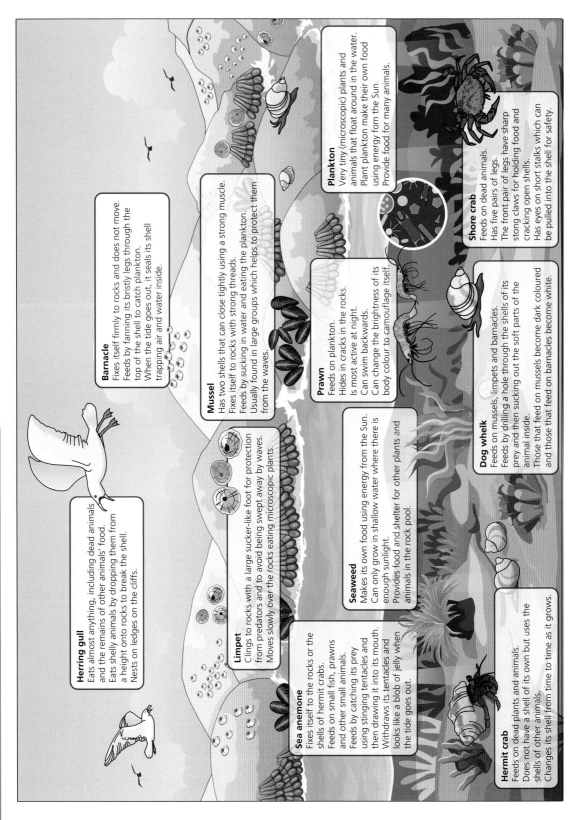

Herring gull
Eats almost anything, including dead animals and the remains of other animals' food.
Eats shelly animals by dropping them from a height onto rocks to break the shell.
Nests on ledges on the cliffs.

Barnacle
Fixes itself firmly to rocks and does not move.
Feeds by fanning its bristly legs through the top of the shell to catch plankton.
When the tide goes out, it seals its shell trapping air and water inside.

Mussel
Has two shells that can close tightly using a strong muscle.
Fixes itself to rocks with strong threads.
Feeds by sucking in water and eating the plankton.
Usually found in large groups which helps to protect them from the waves.

Plankton
Very tiny (microscopic) plants and animals that float around in the water.
Plant plankton make their own food using energy fom the Sun.
Provide food for many animals.

Shore crab
Feeds on dead animals.
Has five pairs of legs.
The front pair of legs have sharp stong claws for holding food and cracking open shells.
Has eyes on short stalks which can be pulled into the shell for safety.

Prawn
Feeds on plankton.
Hides in cracks in the rocks.
Is most active at night.
Can swim backwards.
Can change the brightness of its body colour to camouflage itself.

Limpet
Clings to rocks with a large sucker-like foot for protection from predators and to avoid being swept away by waves.
Moves slowly over the rocks eating microscopic plants.

Seaweed
Makes its own food using energy from the Sun.
Can only grow in shallow water where there is enough sunlight.
Provides food and shelter for other plants and animals in the rock pool.

Dog whelk
Feeds on mussels, limpets and barnacles.
Feeds by drilling a hole through the shells of its prey and then sucking out the soft parts of the animal inside.
Those that feed on mussels become dark coloured and those that feed on barnacles become white.

Sea anemone
Fixes itself to the rocks or the shells of hermit crabs.
Feeds on small fish, prawns and other small animals.
Feeds by catching its prey using stinging tentacles and then drawing it into its mouth.
Withdraws its tentacles and looks like a blob of jelly when the tide goes out.

Hermit crab
Feeds on dead plants and animals.
Does not have a shell of its own but uses the shells of other animals.
Changes its shell from time to time as it grows.

■ Rock pool habitat

Exercise 3.3a

Look at the picture of the rock pool habitat and use the information to help you to answer these questions.

1 The barnacle and the limpet are animals that live in shells. Describe the ways in which they are similar, and the ways in which they are different.

2 The mussel also attaches itself to the rocks – how does it do this?

3 Suggest why these animals have adaptations to help them attach themselves to rocks.

4 What adaptations do sea anemones have to help them catch food?

5 Describe two adaptations of the prawn that help it escape predators.

6 Suggest why seaweeds are not found deep in the ocean.

7 Describe an important difference between a hermit crab and a shore crab.

8 Herring gulls have learnt how to break the hard shells of their prey. Describe how they do this.

Exercise 3.3b

Look at the picture of the rock pool habitat,

Here are some animal facts. Name the animals or plants that are being described.

1 This animal catches its prey with stinging tentacles.

2 These animals fix themselves to the rocks with strong threads.

3 This animal breaks the hard shells of its prey by dropping them onto rocks.

4 These tiny plants are food for many animals in the sea.

5 This animal can change its colour and swim backwards to escape predators.

6 This animal uses the shells of other animals for protection.

Exercise 3.3c: extension

1 Describe in your own words the relationship between the hermit crab and the sea anemone and explain how they benefit each other.

2 Most crabs have hard shells of their own. Hermit crabs have soft bodies and need to use the shells from other animals for protection. Suggest one advantage and one disadvantage of this adaptation.

Investigate a habitat

Look around the area where you live and try to identify some different habitats. Maybe you have woodland, fields or moorland near you. If you live in a town, there will be lots of habitats in your local park or even in your school grounds. Here are some suggestions of habitats for you to study. An identification book or key for the kinds of animals and plants you will find in the habitat will be useful.

Remember that a habitat is the home for a community of plants and animals. Take care of it when you are studying it and try to leave everything as you found it. Remember to wash your hands straight away after doing fieldwork.

Under a log

Try to find an old log rotting on the ground, and carefully lift it up. Very gently lift away pieces of bark and look underneath. Hunt through the soil and any dead leaves that you find around the log.

Try to identify any creatures that you see. Use a soft paintbrush or a spoon to move the creatures gently, and use a hand lens to help you to see the animals more closely.

When you have finished with the log, remember that it must be carefully replaced exactly where you found it. It is a habitat, and the animals must be allowed to return to it.

Shake a tree

Find a leafy bush or tree and spread an old sheet on the ground underneath it. Carefully shake one or two of the branches above the sheet. You need to shake firmly but be careful not to damage the bush or tree. You will find that many of the small animals that live in the bush or tree will fall off onto the sheet.

Use a soft paintbrush or spoon to move the animals to a small container so that you can look at them closely with a hand lens. Try to identify the animals.

When you have finished, tip the animals gently off the sheet under the tree. They will quickly find their way back into the branches.

Pond dipping

Pond dipping is always fun, but remember that water can be dangerous so make sure an adult is nearby if you are dipping in a real pond. An open water butt or rainwater barrel makes a good little pond if you do not have a real pond near you.

Put some water from the pond into a shallow container. A white container is best or a clear one with a sheet of white paper underneath it. Use a small net and move it slowly through the water in the pond.

Gently turn out the animals you find into the shallow container of water so that you can look at what you have caught. Try to identify the animals.

When you have finished, carefully lower the container into the pond so that the animals can be returned gently. Do not pour them back from a height as that could damage or even kill them.

 Pond dipping

Signs of larger animals

The animals you find in the first three activities will mostly be invertebrates. It is harder to study the vertebrate animals in your habitat because they hide or move away when people come near them. You may, however, be able to spot birds. Which birds can you find in your habitat? Can you think of a way to identify birds that you

Deer tracks. Tracks are often the only way we can tell if an animal is present in a habitat

cannot see? Maybe you can tempt them to come a little closer if you put out some food for them. You will need to hide indoors or sit very still when watching them.

Larger wild animals, such as mammals, are much harder to spot. You may see a squirrel in the trees because they are quite bold, but you are less likely to see a deer, a field mouse or a fox. Instead, you need to look for signs that they have been there. Look for tracks in soft ground, nibbled nuts or acorns or maybe bark chewed off the trees. There are many more animals in your habitat than you realise.

Don't forget that the plants in the habitat are also part of the community. How many different species can you find and identify?

How old is the tree?

You can work out the age of a felled tree by counting the annual rings that you see on a sawn cross-section of the trunk. The bark is the outer protective layer of the trunk, and underneath there is a thin layer of living cells called the cambium. Every year these cells divide and make a new layer of wood so the trunk grows outwards, making another annual ring. The tree becomes a little bit thicker and stronger every year. If the weather is dry the annual ring will be narrow. If the weather is wet the ring will usually be wider. You can count the number of rings to estimate the age of the tree.

You can also estimate the age of a tree that is living and growing without sawing it down. Every year, as the cambium makes a new layer of wood, the tree grows a little thicker – about a 2.5 cm increase in girth each year.

1 Measure around the girth of the tree with a tape measure, at about 1 metre above the ground level, and note down the number of centimetres.

2 Using a calculator, divide the girth of your tree by 2.5 and the answer will be the approximate number of years that your tree has been growing.

⇨ **Adaptation**

Why are there so many different organisms?

We have seen that there are many different habitats in the world. We also know that there are millions of different types of organism alive on Earth today. No one really knows exactly how many but it is probably at least 10 million. Fewer than 2 million of these have been studied and given proper scientific names. There is clearly a lot of work for biologists still to do!

In any one habitat, there may be any number of different organisms, depending on the size of the habitat and the opportunities for the organisms to feed, breed and find shelter. This range of different living things is called biodiversity.

But why are there so many living things? One reason is that living things change over time. They become better adapted to life in a particular habitat. Each habitat has different conditions, different challenges and different opportunities and so each may have slightly different varieties of animals and plants living within it. Animals in two or more different habitats may come from the same group but have differences in their body shapes or behaviour in order to cope with the particular conditions in their habitat.

Three foxes

Red fox

You will probably recognise this animal. It is a red fox and is seen almost everywhere in Britain. It is a clever animal and able to adapt its behaviour to allow it to live successfully in a range of habitats. In towns and cities, it is commonly seen in gardens. It will eat fruit, insects and earthworms as well as catching rats, mice and birds. It will also raid dustbins. Like all foxes it has eyes on the front of its head, to allow it to judge distances. This shows us it is a predator.

■ The red fox is commonly seen in towns as well as in the country

In the country, a red fox will catch rabbits, pheasants and small rodents, raid nests for eggs and forage in the hedgerows for fruit. It will take the opportunity to kill and eat chickens or ducks if their owners do not lock them up at night. The red fox is agile and quick, with good eyesight and hearing. Its reddish brown colour seems quite bright when it is out in the open but provides good camouflage in woodland.

Arctic fox

This is a different species of fox. It lives in cold areas in the far north of Europe and North America where temperatures can drop to −50 °C or lower in winter. This fox is a little smaller and stockier than the red fox and its coat is much thicker. The Arctic fox's fur consists of two different types of hair. The under fur is thick and soft. It traps air and acts as a good insulator. Over the top of this is a layer of coarser, stronger fur that is waterproof.

■ The Arctic fox is well adapted to very cold conditions

The Arctic fox's winter fur is the warmest fur of any animal, even warmer than that of the polar bear. It also has fur on the underside of its feet which helps it walk on icy surfaces as well as keeping its paws warm. Extra blood vessels in the feet help to stop the fox's feet from freezing and sticking to the cold ice. It has a long, bushy tail, which it uses to reduce heat loss by covering its nose and face as it curls up in a snow hollow during stormy weather. The Arctic fox has small ears and short legs, because these are areas where heat can be lost easily. It also has a thick layer of body fat.

The fur of the Arctic fox is pure white in winter when the ground is covered with snow but becomes mottled with brown when the snow melts in the short Arctic summer. This great camouflage helps it to get close to its prey without being seen.

Arctic foxes eat a lot of small rodents called lemmings. In summer, lemmings often breed very successfully and the foxes have a lot to eat. In the winter the lemmings spend most of their time in tunnels in the snow so the foxes need their very good hearing and sense of smell to find them. If the fox manages to kill more than it can eat, it will bury the extra in the snow or in the frozen ground. The food will stay quite fresh until the fox wants to eat it, just as food keeps fresh in the fridge or freezer at home. Like their red cousins, Arctic foxes are quite adaptable in their behaviour. If there are no lemmings around they will eat seal pups, scavenge on the bodies of animals that die in the harsh weather or eat the remains of a polar bear kill.

Fennec fox

Fennec foxes live in the deserts of northern Africa and the Middle East where it is very hot and dry. They are tiny and are the smallest members of the fox family, but they have huge ears that look as if they belong to a different animal entirely! These massive ears allow excess heat to travel out of the fox's body and into the air, keeping the fox cool.

■ The fennec fox's large ears help it to stay cool

Fennec foxes have quite thick fur. This keeps them cool in the hot daytime and warm at night when it gets cold. They have fur on their feet to stop them burning themselves on the hot sand. This fur also gives the feet a larger surface area and acts like a snowshoe, helping to stop the fox sinking into the soft sand.

Like the other foxes, fennec foxes will eat whatever they can find. This may include small rodents, lizards, insects, birds' eggs and plant material. Because there is so little water available in their habitat, they have developed the ability to survive for long periods without drinking.

All three foxes are well adapted to living in their particular habitat. Each of the habitats is very different and so, although the foxes are all from the same family and have many similarities, they also have some important differences.

Two plants

Bluebells

Here is a plant you may recognise. It is our native bluebell and is found in deciduous woodlands.

◾ Bluebells flower early, before the leaves of the woodland trees are fully open

Bluebell leaves and flowers are produced from bulbs found underground. In the spring, as soon as the weather begins to warm up a little, the bluebell leaves begin to push up through the soil. They grow very quickly and soon the blue flowers open and carpet the floor of the woodland. The attractive blue petals on the flowers indicate they are insect-pollinated, but there are not very many insects around early in the year, and a late frost can damage or kill the flowers. So, why do bluebells flower so early?

The woodland trees, forming the canopy above the bluebells, are deciduous and lose their leaves in the winter. When there are no leaves on the trees, light can reach the floor of the wood more easily. Plants growing there can absorb enough light to carry out photosynthesis and make a lot of food. Some of this food gives them the energy to grow and to make flowers and the rest is stored in the bulbs under the ground.

Soon after the bluebell flowers appear, the leaves begin to grow on the trees. The light levels on the floor of the wood gradually decrease, which means there is less and less light energy for the plants to carry out photosynthesis. The bluebell must reproduce quickly in the short time before this happens. The flowers are sweetly scented and contain nectar to attract the few insects.

After fertilisation, lots of seeds form in rounded seed cases. The leaves of the plant then begin to die and, after a very short time, you would never know that there had been any bluebells at all. All through the summer, autumn and winter, the plant is hidden underground as a bulb packed with food to help it grow new leaves the following spring.

But what happens if there are no insects one year or the flowers are damaged by frost so they cannot make seeds? The bluebell plant has another trick to help it to reproduce. It can make tiny little bulbs attached to the main bulb under the ground. These gradually grow and, after a year or two, can break away from the parent bulb and become a separate plant. The bluebell is clearly well adapted to the conditions under the trees in deciduous woodland.

■ The bluebell stores food in the bulb to help it survive

Cactus

A cactus is very different to a bluebell. To start with, where are its leaves? It is fat and prickly and not at all like a typical plant.

Cacti are adapted to life in very hot, dry desert conditions. In this environment normal leaves would quickly dry and shrivel up, and then not be able to carry out photosynthesis. Cactus leaves have adapted to become hard, sharp spines, which protect the plant from hungry, thirsty animals that might want to eat it. They do not contain chlorophyll and no longer carry out photosynthesis to make food for the plant. Instead, the cactus stem has taken on the role of food production from the leaves. It has become a big water reservoir and, because it is so large it has a bigger surface area than most plant stems, which allows it to absorb enough light for photosynthesis.

■ This cactus's stem is a water store and the leaves have become spines

The shape of the cactus is also an important adaptation. If there is any rain or dew, it collects on the surface of the plant. The cactus's shape makes sure that any water runs down to the roots where it can be taken up into the water store in the stem. To make sure the valuable water is not lost through evaporation, the cactus is covered in a very thick waxy waterproof layer, which is also quite tough for animals to eat.

Exercise 3.4a

1 Look back at the section called 'Three foxes'.

  (a) List as many similarities as you can between the three foxes' bodies and their behaviour.

  (b) List as many differences as you can identify between the Arctic fox and the red fox.

  (c) Why would the fennec fox not be able to survive in the Arctic?

2 Look back at the section called 'Two plants'.

  (a) Why do bluebells flower early in the spring?

  (b) What problems might there be when flowering early in the year?

  (c) How do the bluebell plants survive during summer, autumn and winter when they have no leaves to make food?

  (d) Describe in your own words how a cactus is adapted to survive in the desert.

Exercise 3.4b

Use the words below to complete the sentences. Each word may be used once, more than once or not at all.

chickens   dustbins   ears   eaten   fruit   leaves

lemmings   sunlight   predators   prey   rabbits

spines   three   two   water

1 Foxes have eyes on the front of their heads because they are

  _____ .

2 Red foxes that live in towns will often raid _____ to get food.

3 In the country, red foxes may eat _____ , _____ or pheasants.

4 Arctic foxes have _____ types of hair to trap heat and keep them dry.

5 Arctic foxes eat a lot of _____.

6 Fennec foxes have big _____ to take heat away from their bodies and keep them cool.

7 Bluebells flower early before the trees grow _____ which block out the _____.

8 The leaves of cacti have changed to become _____ to protect the plant from being _____.

9 Cacti have large stems to store _____.

Exercise 3.4c: extension

Read the following description of a camel.

The best-known desert animal is probably the camel. It is extremely well adapted to desert life, which is why the people living there rely on it for transport across the desert sands. The type of camel most often found in the North African Sahara desert is the one-humped dromedary camel. The camel's hump is an important adaptation. It is not used to store water as many people believe. Instead it is full of fat, which provides the camel with a food store, helping it to survive many days travelling between oases (watering holes). When a camel comes to a pool of water, it can drink up to 100 litres in one go and then store it in its large stomach.

To save water and prevent themselves from becoming dehydrated, camels do not sweat unless their bodies warm up to over 40°C. Most animals would die if they got this hot. Camels also lose less water through their nostrils than other animals. Their long legs keep their bodies further from the hot sand and their thick fur reflects heat away from the body to keep them cool.

Camels are able to close their nostrils and have very long eyelashes and hairs inside their ears. These adaptations are very useful in sandstorms because they stop the sand blowing into their noses, eyes and ears. Their large flat feet are ideal for walking across soft sand.

1 You need a picture of a camel in the centre of a sheet of paper. Your teacher may give you one of these or you could draw it yourself. Label the diagram neatly, clearly showing the ways in which the camel is adapted to life in the desert.

2 Make up at least five questions about the camel. Make sure that the answers to these questions can be found in the description you read above. Write each question in your book and then write a model answer to it, including as much detail as possible. See if your partner can answer the questions.

3 Think of one or two questions about camels that cannot be answered by reading the description. Write them down and try to find the answers from books or the internet.

Camouflage

One of the most obvious ways in which an animal is adapted to its habitat is its colouring, otherwise known as its camouflage. Camouflage helps animals to hide. Prey animals need to hide from predators. Predators need to hide so that their prey does not see them lying in wait or approaching. Even some plants are camouflaged. For instance, the stone plant, found in stony deserts, looks exactly like the stones that surround it. It is very hard to spot.

■ Stone plants are hard to see amongst pebbles

Some animals are so well camouflaged that it is hard to see them even when you know they are there. This bird, a nightjar, which keeps its eyes almost shut, looks just like some fallen leaves or a dead twig and could easily be overlooked.

The animals in the following pictures are also well camouflaged. They are both insects. You may be able to spot them and guess what they are pretending to be!

■ Can you spot the nightjar?

■ These insects are very well camouflaged

4 Properties of materials

⇨ What are materials?

What do you think of when someone uses the word 'material'? To many people, materials are what we use to make our clothes, for example cotton, wool or polyester. To a scientist, materials are what we use to make all sorts of things.

There are lots of different types of materials. Wood is used to make tables and doors. Steel is used to make bicycles and cars. Aeroplanes and drinks cans are made from aluminium and lots of things are made from plastic. Wood, steel, aluminium and plastic are all materials.

Properties of materials

How do we decide what materials to make things from? Why is wood a good material to use when making a table? Would a wooden saucepan be a good idea?

To decide which material to choose, we need to know about how materials behave in certain conditions. A table needs to be strong and reasonably easy to make. Wood is a strong material that can be easily cut and shaped. Wood is a good material to use for making tables. However, wood also burns easily and does not allow heat to pass through it very well so it would not be a sensible choice for making saucepans. You have already learnt that some materials are magnetic but most are non-magnetic. You have also learnt that metals are good electrical conductors but most other non-metals, such as plastic, are electrical insulators.

The ways in which a material behaves are called its properties. Each material has several properties and it is important to know about all of them to make the right choice. Often, more than one property is important when making something. The material used to make a raincoat needs to be waterproof, but it also needs to be flexible and hard-wearing and to feel comfortable.

Did you know?

Often materials can be recycled and used again. For example, some fleece fabrics are made from recycled plastic cups. You will learn more about plastics and recycling in Chapter 7.

Activity – which material?

Take a look at these pictures. Which materials are used to make these things?

Could the objects in the pictures have been made from any other materials? Work with a partner or in a group and see how many other materials could have been used.

Activity – spot the material

Look around the room. How many different materials can you find?

Make a list of all the materials and the things that are made from each material.

Count how many different things are made from each material and then draw a pictogram, like the one to the right, to show your results.

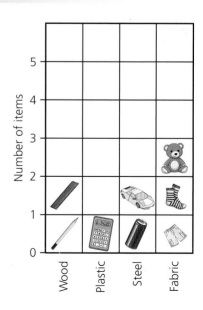

Exercise 4.1a

1 What do scientists mean by the word 'material'?

2 What is the meaning of the word 'properties'?

3 What material was used to make the chair you are sitting on?

4 Suggest some other materials that could be used to make a chair.

5 What properties are important when choosing a material to make a toy for a young child?

6 We usually use glass to make windows. What property does glass have that make it a good material for this job?

7 (a) Suggest another material that could be used instead of glass to make windows.

 (b) What would be the advantages and disadvantages of using this material to make windows?

Exercise 4.1b

Use the words below to help you to answer the following questions. Each word may be used once, more than once or not at all.

break   glass   light   materials   opaque   plastic
properties   transparent   wood

1 A word that means 'what objects are made from' is

 _____.

2 The _____ of a material are the ways in which it behaves.

3 Windows are usually made from _____ because it is _____.

4 A chair could be made from _____ or

 _____.

5 A child's toy might be made from _____ because it is _____ and does not _____ easily.

1 Here is a list of materials that you might find around the classroom. They have been sorted into two groups.

Group 1: wood   steel   glass   brick   thick plastic

Group 2: cotton   thin plastic   carpet   paper

(a) What property of the materials do you think has been used to sort them into these two groups?

(b) Can you think of another way of sorting the same materials?

2 Buckets are usually made from plastic but in olden times they were made from metal, wood or leather. Why do people choose to use plastic now rather than these other materials? Think of as many reasons as you can.

Activity – a properties matching game

Here are some words that are used to describe the properties of materials.

hard   soft   strong   weak   brittle   flexible   rigid   flammable   magnetic   non-magnetic   transparent   opaque   absorbent   translucent

First check that you know what they all mean. Use a dictionary to look up the meanings of any that you do not know.

Now write each word on a separate piece of paper or card. Each group will need one set of cards.

Put all the cards on the table, face down so you cannot read them, and mix them up.

Take it in turns to select a card, read the word on it and then find an object in the room made from a material with that property. Put the card onto the object. For example, you might choose to put 'strong' onto a bookcase because bookcases have to be strong to support the weight of all the books.

Activity – chocolate bicycles

Bicycles are usually made from steel. Steel is strong and hard wearing. Steel is a good choice of material for a bicycle. What would a chocolate bicycle be like? Write a story about someone who has a bicycle made from chocolate.

Testing materials

Activity – finding the properties of materials

Some properties of materials are obvious. For example, you can easily see that glass is transparent, and you can feel that a piece of wood is rigid but a piece of polythene is flexible.

Sometimes the properties are not so obvious and we have to test the material to find out what they are.

In this exercise you are going to test a range of materials to find out about some of their properties.

You will need:

● some materials (e.g. woolly or fleece fabric, different types of metal, plastic, wood, modelling clay/play-dough)

● a magnet

● a circuit with a lamp and a gap in it.

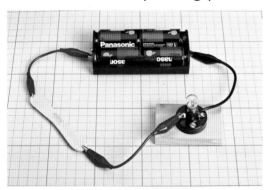

You are going to find out about the following properties:

● electrical conductivity

● thermal conductivity

● magnetic behaviour.

Start by drawing up a table to record your results. It should look like this.

Material	Electrical conductivity	Thermal conductivity	Magnetic behaviour

Test 1: Electrical conductivity
Take each material and place it across the gap in the circuit so that it is touching both of the loose ends or crocodile clips. Does the lamp light up?

If the lamp lights up, the material has allowed the electricity to flow through it to complete the circuit. We call such a material an electrical conductor.

If the lamp does not light, the material has stopped the electricity from flowing through. We call such a material an electrical insulator.

Write *electrical conductor* or *electrical insulator* in the first column of the table to record the electrical conductivity of each material.

Discuss with your partner: Can you suggest one situation where it would be important to use a material that is a good electrical conductor, and one situation where it would be important for the material to be a good electrical insulator?

Test 2: Thermal conductivity
There are lots of ways of testing for thermal conductivity and you will learn more about this in Year 6.

To judge quickly whether a material is a good thermal insulator, all you need to do is to touch it. If it feels cold to the touch, it is carrying thermal energy (heat) away from your hand and is a good thermal conductor.

If the material does not feel cold it means that the energy is not being conducted away from your hand so the material is a good thermal insulator.

Write *thermal conductor* or *thermal insulator* in the second column of the table to record the thermal conductivity of each material.

Discuss with your partner: Can you suggest one situation where it would be important to use a material that is a good thermal conductor, and one situation where it would be important for the material to be a good thermal insulator?

Test 3: Magnetic behaviour
Bring the magnet close to each material in turn. If the material is attracted towards the magnet, we say that it is magnetic.

If the material is not attracted to the magnet, we say that it is non-magnetic.

Write *magnetic* or *non-magnetic* in the third column of the table to record the magnetic behaviour of each material.

Discuss with your partner: Are any of the materials electrical and thermal conductors? Do these have anything in common? Which of these is also magnetic?

Activity – comparing materials

When we make an object we need to choose the best material for making it. Scientists need to compare the properties of materials to find out which one would be best for a particular job. To do this, they need to carry out a fair test.

These scientists are testing materials to see which are suitable for making swim shorts

Planning an experiment

Read the picture story about some people carrying out an investigation. Then discuss with your partner or group whether you think that the people in the story are working scientifically. What kind of investigation are they carrying out? How can you tell?

Now work with your partner or group to carry out the following investigation of your own.

The stretchiest tights

Rebecca and Alice are sisters. They both go to the same school. In the winter they wear tights. One morning, Rebecca took some tights out of her cupboard and started to put them on but they were too small. Alice put hers on but they were too big.

Rebecca thought that it would be good if tights were made of very stretchy material. Then it wouldn't matter if she had Alice's tights by mistake.

How could you find out which tights are the stretchiest? Plan an investigation with your group. It will need to be a fair test so remember to think about all the important variables and what you will need to do to make it fair.

Think about what you will measure and how you will measure it. What apparatus will you need? What will you do to make sure that your results are reliable?

How will you record and report your results? Remember that you need to show them in a way that is clear and easy to understand so you will need to think about how you will do this.

Make a collection of different types of old tights. It is best if they do not have big holes in them for this experiment. Carry out your test to see which ones are made from the stretchiest material.

When you have done your experiments, discuss with your group how well you think you have done this task. Can you suggest any way that you could have improved it?

Exercise 4.2a

1 Some children wanted to find out which type of kitchen paper soaked up water the best. They cut a strip of each paper and stuck them onto a piece of wood.

They dipped the ends of the paper into some coloured water and held them there for a few minutes. Then they took them out and looked at how far the water had risen up the paper.

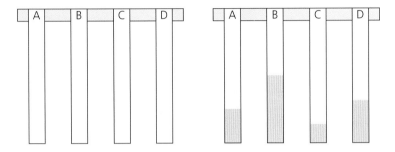

(a) Measure the height of the coloured water mark carefully on each strip of paper. Write your measurements down neatly in your book.

(b) Which paper soaked the water up the best?

(c) Which paper was least absorbent?

(d) Do you think that the children carried out a fair test? Explain your answer.

2 Many people think that we should not use plastic bags because they are not good for the environment. What other materials could be used to make a container to carry shopping home?

Exercise 4.2b: extension

Paper towels need to be absorbent. It is also important that they stay quite strong when they are wet. Describe how you could carry out a test to find out which type of paper towel was the strongest when wet. Remember to explain how you would make it a fair test.

A lucky walk

One day in the summer of 1948, a Swiss man called George de Mestral decided to take his dog for a walk. There were lots of mountains near his home and George enjoyed walking and climbing. George and his dog walked for a long time up and down the hills. The dog ran back and forth through the grass and undergrowth as they went.

When they came home, George noticed that his dog's fur was full of little seed cases called burrs. They had come from burdock plants and were hard to remove. George, who loved inventing things, wondered how it was that the burrs held onto the dog's fur so tightly. He looked down and discovered that there were burrs stuck to his trousers too.

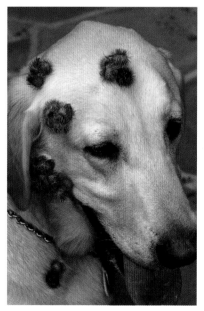

■ The burrs are attached to the dog's fur

When he had finished grooming the dog, George took one of the burrs and looked at it with a microscope. He discovered that the burr had lots of tiny hooks on it. These hooks tangled themselves up in the dog's fur and held on tightly. He knew that this was to help the plant to disperse its seeds.

George decided that he could use this discovery to make two tapes that would hold onto each other when they were pushed together. One tape would have little tiny hooks on it and the other would have little loops. These tapes, when sewn onto clothes or other objects, would work like a zip.

■ Burrs are covered with tiny hooks

He told his friends about his idea. They all laughed. They thought it was a rather silly idea but George was sure it would work. He went to visit a weaver he knew and they worked together to make the new hook and loop fasteners. They tried different materials to make the hooks and different ways of weaving the loops. At last, in 1955 after a lot of trial and error, they succeeded in making their new fastener work. They called it Velcro® from the French words for 'velvet' and 'hook'.

Now we find Velcro® on all sorts of things. Maybe you have a jacket, a schoolbag or even shoes with Velcro® on them. It is quick and easy to use and George saw his invention in use all over the world before he died in 1990.

■ Velcro® is now used all over the world

Exercise 4.3a

1 What activities did George de Mestral enjoy in Switzerland?

2 What plant did the seeds George de Mestral found in his dog's fur come from?

3 How had they become attached to the fur?

4 Explain how this might help the plant.

5 Describe the idea that George de Mestral had after looking at these seeds.

6 How did his friends react to his idea?

7 Why did he call his invention Velcro®?

Exercise 4.3b

Fill in the gaps in the following sentences using the words below.

burrs   hooks   loops   Switzerland   Sweden   Velcro®
velvet

1 George de Mestral lived in _____ .

2 George discovered seed cases called _____ stuck to his dog's fur.

3 He saw that these seed cases stuck to the fur because they were covered in _____ .

4 He made special tapes with hooks and _____ on them to act like a zipper.

5 He called his invention _____ from the French words for _____ and hook.

Exercise 4.3c: extension

1 Young children often have Velcro® on their shoes because it is easier to fasten than laces or buckles. Think of other places where Velcro® is used and describe what might have been used if Velcro® had not been invented.

2 Any new product needs to be advertised to let people know about it and show them why they might find it useful. Imagine that you have been asked to promote George de Mestral's new invention and design an advertisement. It should be eye-catching and informative.

5 Reversible changes

⇨ Changes of state

You have already learnt that different materials have different properties. One of these properties is the way that the material behaves when it is heated or cooled. Often heating or cooling results in a change of state. Can you remember the names of the changes of state? Look carefully at the pictures below and discuss with your partner or group what name is given to the change of state being shown.

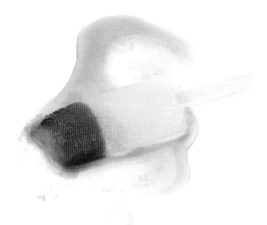

■ An ice lolly will turn from a solid to a liquid as it warms up

■ The water in these wet clothes will turn into a gas if it is warm and windy

■ The water in this tray will turn into a solid when it gets really cold

■ When the water vapour in the air meets a cold surface it turns into droplets of liquid water

We often describe these changes as reversible. This means that the substance can change between states without becoming a different substance. This property is very important in many ways and you will learn more about these later in this book.

Did you know?

Water is a very unusual liquid. When water freezes it expands. Most liquids will contract when they freeze. This is good news for plants and animals that live in ponds because a layer of ice will float on top of the water and help to stop the whole pond from freezing. However, it is bad news if water pipes in your home freeze, because the ice can split the pipes as it freezes. When it begins to melt, the water will drip out through the split. If water freezes in cracks in rocks, it can split the rocks apart. We call this physical weathering and you may have learnt about it in your geography lessons.

■ A burst water pipe can cause flooding in the house

⇨ Dissolving – another reversible change

If you stir a little sugar into a cup of tea, the sugar seems to disappear. It can no longer be scooped out of the cup. However, when you drink the tea you can taste the sugar so it is still in the cup. What has happened to the sugar? You may hear people saying that the sugar has melted in the tea. In fact this is not the case. Melting is when a substance is heated and turns from a solid to a liquid. Something different happens when the sugar is mixed with the water.

Like all materials, sugar crystals are made of particles. When the sugar is stirred into the tea, the crystals break apart and the sugar particles mix with the liquid tea particles. We say that the sugar has dissolved.

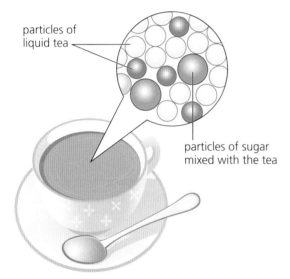

particles of liquid tea

particles of sugar mixed with the tea

■ The sugar dissolves in the tea

Substances like sugar that dissolve are described as soluble.

Not all materials are soluble. Some materials cannot dissolve and are described as insoluble. Sand, for example, is insoluble, which is just as well because otherwise it would dissolve into the sea and there would be no beaches to make sand castles on!

Activity – soluble or insoluble?

You will need:

- beakers
- warm water
- a spoon or a spatula
- stirring rod.

Your teacher will give you some materials to test, for example, salt, pepper, flour and sawdust.

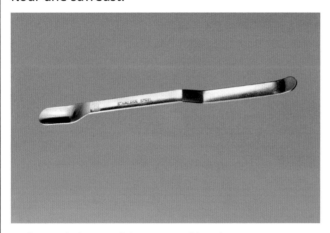

■ A spatula is a special scoop used in science

Half fill a beaker with water and add a little of one of the materials using the spoon or spatula. Then give the mixture a good long stir.

Look carefully at the mixture and describe clearly what you see.

Leave the mixture for a while and then take another look. Does it look any different? If so, describe what it looks like now.

If the mixture becomes a clear liquid, the material has dissolved and it is soluble. If the material does not disappear, the material is insoluble.

Sometimes the material is made from such tiny pieces they float around in the mixture for a while, making the mixture appear opaque or milky. This mixture is called a suspension. After a time, the tiny pieces will drift down towards the bottom of the beaker and you will see them at the bottom of the beaker with clear water above. These substances are also insoluble.

Decide whether the materials you tested are soluble or insoluble.

Many of the key words you will come across in this chapter begin with the letter 'S'. When a soluble substance dissolves, we call the mixture a solution. A solution is made of two substances.

The solid that dissolves is called the solute.

The liquid it is dissolved in is called the solvent.

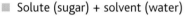

— solvent

sugar lumps

■ Solute (sugar) + solvent (water)

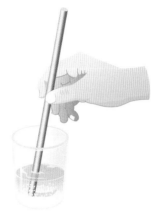

■ makes a solution (sugar syrup)

Exercise 5.1

Use the words below to complete the sentences. Each word may be used once, more than once or not at all.

can	cannot	dissolved
insoluble	soluble	solute
solvent	suspension	solution

1 When a substance seems to disappear when mixed with water, we say that it has _____.

2 A substance that is able to dissolve is described as a _____ substance.

3 An insoluble material is one that _____ dissolve.

4 When a substance dissolves in water, the mixture is called a _____.

5 In a solution, the liquid part is called the _____ and the dissolved solid is called the _____.

6 A _____ is formed if the pieces of an insoluble substance are tiny enough to float in the liquid.

Investigating changes in mass

In this activity it is really important that you measure masses and volumes very accurately so take care when measuring and make sure that you do not spill anything.

You will need:

- a conical flask
- a measuring cylinder
- scales
- sugar
- a small container, for example a petri dish, to weigh the sugar into.

Take a flask and find its mass. Remember to write down the mass clearly, including the units, in your notebook.

Measure out 50 ml of warm water, using a measuring cylinder, and pour it carefully into the flask, making sure that you do not spill any. Weigh the flask again. You now have the mass of the flask and the water together.

Use the two masses (*empty flask* and *flask + water*) to calculate the mass of the water by itself.

Carefully weigh out 2 g of sugar and record its mass.

Discuss with your partner the following question:

If you weigh the flask and its contents again after you have added the sugar, what will the mass be?

Carefully pour the sugar into the flask and swirl it round gently to dissolve the sugar. Do not use a stirring rod. Take care not to spill any of the mixture.

Weigh the flask and its contents. Was your prediction correct? Discuss your findings with your partner or group to see if you can explain your observations using what you know about particles. Share your ideas with the class to see if they all agree.

Now discuss the following questions:

1 What was the mass of the water you put into the flask?

2 What was the mass of the sugar that you added to the flask?

3 What is the mass of the solution? (Hint: you will need to work this out)

4 What do you notice about these figures?

If your measurements have been accurate you should have discovered that the mass of the solution is equal to the mass of the solvent plus the mass of the solute. This is always the case when we make a solution and we call it conservation of mass.

Here is a tricky extra question to think about. Look back at the measurements you made at the start of the activity above. Now discuss whether you could predict the mass of the solution if you knew the volume of the water and the mass of the solute.

Making things dissolve more quickly

People are often in a hurry. We do not like to wait for ages for the sugar to dissolve in our tea and so it is useful to think about how we can make it happen more rapidly.

Speeding up dissolving

Working Scientifically

How can we make things dissolve more quickly? Discuss your ideas with your partner or group. Here are some questions to help your discussion.

- When we put sugar in our tea, we always stir it. Does this make the sugar dissolve more quickly?

- Does sugar dissolve more quickly in hot or cold liquids?

- Sometimes people have special sugar with big crystals for putting into coffee. Does this sugar dissolve more quickly or more slowly than ordinary sugar?

Activity – sugar racing

The questions above mention three variables (factors) that might affect how quickly sugar dissolves:

- stirring

- the temperature of the water

- the size of the sugar crystals.

Choose one of these variables (factors) and plan an experiment to test whether it affects how quickly sugar dissolves.

Remember to make your investigation a fair test. Record your results neatly in a table.

When you have all done your testing, discuss the results with the rest of your class so that you all have the answers to all the questions.

■ Remember to work safely when doing chemistry experiments

Exercise 5.2

1 Copper sulfate is a blue chemical that dissolves in water to make a blue solution. In copper sulfate solution, what is:

   (a) the solvent

   (b) the solute?

2 A scientist wishes to make a solution of copper sulfate quickly. Name three things she should do to help her do this.

3 What is a suspension?

4 Some pupils want to find out whether the temperature of the water affects how quickly copper sulfate crystals dissolve.

   (a) What is the one variable (factor) they should change in their investigation?

   (b) Which variables (factors) should they keep the same to make their test fair?

   (c) What pattern would you expect them to find in their results?

Saturated solutions

You have learnt that a solution is made by dissolving a solute (soluble substance) into a solvent (such as water). If you go on adding more and more of the solute you will find, after a while, that no more of the solute will dissolve, however much you stir. A solution like this, where no more solute will dissolve, is described as a saturated solution. Some substances, such as sugar, are very soluble and a lot of sugar must be added before you make a saturated solution. Other substances are less soluble and so less needs to be added before the solution is saturated. Examples of less soluble substances are salt and copper sulfate.

You have already found that substances will dissolve more quickly in hot water. Many substances are also more soluble in hot water so the quantity of solute needed to make a saturated solution may depend on the temperature of the solvent.

Activity – growing crystals

You will need:

- a beaker containing 100 cm³ hot water (be very careful with the hot water!)
- another clean beaker
- fine copper sulfate crystals
- a spatula
- a stirring rod
- one larger copper sulfate crystal
- thread
- a pencil or wooden splint
- eye protection.

Remember to wear eye protection all the time while you are doing this experiment.

Add fine copper sulfate crystals to the hot water carefully, using the spatula.

Stir well after each addition to dissolve the crystals. You will notice that the colour of the solution becomes darker and darker.

When you are sure no more crystals will dissolve, leave the solution to cool to room temperature. Wash your hands thoroughly before you do anything else.

When it has cooled, pour the solution carefully into a clean beaker. You will probably have some crystals at the bottom of your beaker. Try to leave these behind. Your teacher may give you a filter to do this. We will learn more about filtering in Chapter 6.

You now have a saturated solution of copper sulfate.

Carefully tie the thread around the larger copper sulfate crystal and wrap the other end of the thread around the pencil or splint.

Balance the pencil or splint across the top of the beaker and adjust the length of the thread until the crystal is hanging in the solution and not touching the bottom of the beaker. This crystal is called a seed crystal.

Put the beaker in a safe place and then wash your hands before removing your eye protection.

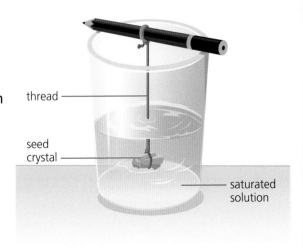

Leave your solution for a week or two, checking it from time to time. Your crystal should grow bigger and bigger. This happens because the water in the solution slowly evaporates and so less and less copper sulfate can stay dissolved. The extra copper sulfate sticks to your crystal and makes it bigger.

thread

seed crystal

saturated solution

What shape is your crystal? Is it the same shape as the ones made by other members of your class?

You might like to try making crystals of other substances, such as salt or sugar.

Exercise 5.3a

Some pupils carried out an experiment to find out how the temperature of the water affects the amount of the solute needed to make a saturated solution. They did the experiment twice, using a different solute each time.

Here are their results:

Temperature of the water in °C	Quantity of solute that can be dissolved in 100 cm³ of water in grams	
	Solute A	Solute B
20	36.0	21.0
40	36.5	29.0
60	37.0	40.0
80	38.0	56.0

1 Describe how the temperature of the water affected the amount of solute B that would dissolve in 100 cm³ of water.

2 The pupils' teacher told them that the solubility of salt does not change very much with temperature. Which solute might have been salt?

3 Suggest how much of solute A might dissolve in water at 100 °C.

4 Draw a bar chart to show the results for solute B.

Exercise 5.3b: extension

1 On graph paper, use the data from Exercise 5.3a to draw a line graph to show both sets of results. Put 'Temperature of water in °C' on the horizontal axis and 'Quantity of solute in grams' on the vertical axis. Plot the points for solute A first and then join them carefully with a smooth curve. Then plot the results for Solute B and join these points with another smooth curve. You might like to use a different coloured line for each solute. Remember to label the lines or use a key to show which line is which. Give your graph a suitable title.

2 Use your graph to find the following information:

(a) At what temperature is the solubility of solute A the same as solute B?

(b) How much of solute B would you be able to dissolve in 100 cm³ of water at a temperature of 30°C?

(c) At what temperature would you be able to dissolve 50 g of solute B in 100 cm³ of water?

3 Suggest how much of solute A would dissolve in 200 cm³ of water at a temperature of 60°C.

Separating mixtures

6

In the last chapter we learnt that physical changes in substances, such as changes of state and dissolving, are reversible. This fact can be very useful in science, for example when we have a mixture of substances that we need to separate.

⇨ Mixtures are all around us

What is a mixture? A mixture is just two or more things mixed together. For scientists, however, mixtures are very important and can pose some interesting scientific challenges.

Let's start by thinking about mixtures you may be familiar with. Sometimes it is easy to see that things are mixed together. For example, the grains, nuts and fruit in a bowl of muesli can be seen clearly and it would be quite easy to pick out each of the different foods to separate the mixture.

Sometimes, though, it is not possible to see the individual components of the mixture. For example, it is not possible to see the sugar in a sugar solution. In fact, the mixture looks like pure water.

Air is a mixture of gases. Most of the air is a gas called nitrogen and it is mixed with oxygen. It also contains much smaller amounts of carbon dioxide, water vapour and other gases, for example, argon.

Sea water is a mixture of water and several different salts. Among these are common or table salt, which chemists call sodium chloride, and which makes the water taste salty. The water also contains microscopic animal and plant material called plankton and possibly mud, sand, shell fragments and bits of seaweed. Often sea water is polluted with material that we have added to it. This could be waste from factories, chemicals washed into the sea from farmland or even sewage.

All of these are mixtures. Now let's think scientifically about them. Do you think the mixture of gases that makes up the air is always the same? What might change the mixture? For example, is the air you breathe in the same as the air you breathe out? In what ways might it be different? Can you think of any other ways in which the mixture that we call air might change? What about sea water? Is all sea water the same?

Did you know?

The water in the Dead Sea, a huge salt water lake between Israel and Jordan, is more than eight times as salty as the big oceans. The huge amount of salt makes the water very dense and people can float easily in the water. The Dead Sea is the lowest place on Earth at roughly 400 metres below sea level.

As we have already stated, a mixture is made up of two or more things mixed together, but the really interesting thing about mixtures is that they can change but still be mixtures. For example, the air we breathe out contains less oxygen, more carbon dioxide and more water vapour than the air we breathe in but we happily call both of them 'air'.

Activity – think of some more mixtures

Work with a partner or in a small group. See how many examples of mixtures you can think of and write them down. Then think about how these mixtures might change or be changed by natural events or by human actions.

You should be able to think of quite a number of mixtures, as pure substances are quite unusual. Almost everything contains at least a little bit of something else, either deliberately or by chance. For example, you may think that the water you drink is pure, but it almost certainly isn't. Our drinking water contains dissolved chemicals that give it some taste, so it is probably a good thing that the water is not pure. If you tasted water from different parts of the country, or maybe different countries, you might be able to taste the difference.

Activity – investigate different waters

Water from different places is different. Your teacher will give you some water samples to investigate. You may have some sea water, pond water, tap water, distilled (pure) water or some mineral water.

For this activity you will need:

- a sheet of rigid clear plastic
- a sheet of black paper
- water samples
- pipettes (one for each water sample)
- sticky labels (one for each water sample).

Place the sheet of plastic on top of the sheet of black paper in a place that is warm but not windy and where no one will knock it by mistake.

Use the pipettes to put a little 'puddle', about 1 cm across, of each water onto your sheet of plastic and label each puddle.

Try to make all the puddles the same size. Then leave your sheet alone until all the water has dried up (evaporated).

Explain why it is important to have one pipette for each type of water.

Try to predict what you will find when the water has dried up.

Explain your ideas to your partner. Record your predictions and final observations in a table like the one below:

Water sample	Prediction	Final observation

1 What is a mixture?

2 Name some of the gases that make up the mixture we call air.

3 Is all air the same? Explain your answer.

4 What materials may be present in the mixture we call sea water?

5 Describe how you could show that sea water contains more dissolved materials than tap water.

6 What is the chemical name used by chemists for table salt?

7 Where is the Dead Sea?

8 Why is it especially easy to float in the Dead Sea?

9 What is special about the position of the Dead Sea?

Use the words below to complete the following sentences. Each word may be used once, more than once or not at all.

| distilled | gases | highest | lowest | mixture |
| nitrogen | oxygen | pure | salt | sodium |

1 A _____ is made up of two or more things mixed together.

2 Air is a mixture of _____ .

3 Most of the air is made from _____ and this gas is mixed with _____ and some other gases.

4 The name that chemists give to table salt is _____ chloride.

5 Our drinking water is not _____ water because it contains dissolved salts.

6 The Dead Sea is the _____ place on Earth.

7 It is easy to float in the Dead Sea because it has so much _____ dissolved in the water.

⇨ Separating mixtures

Scientists often need to separate different parts of a mixture. Chemists often want to obtain pure chemicals for their experiments. These chemicals almost always start as part of a mixture, so we need to be able to separate the mixtures to obtain pure substances. There are many other reasons why people might want to separate a mixture and there are lots of different separation methods. We are going to learn about some of them in this chapter.

Decanting

Decanting is probably the simplest way of separating a mixture. It works for a mixture containing a liquid and pieces of a solid that will sink quickly to the bottom of the container. You decant the liquid by pouring it carefully into another container, leaving the solid material, sometimes called a sediment behind. Pouring can be made easier by holding a stirring rod by the spout of the beaker, so the liquid trickles down the stirring rod rather than dribbling down the side of the beaker and making a mess.

■ A stirring rod makes it easier to pour without making a mess

Did you know?

Sometimes bottles of wine contain solid materials that would not be very pleasant to drink. If the wine is left to stand for a while the sediment sinks to the bottom and then the wine can be decanted into another container. These containers are often called decanters.

■ Wine can be poured into a decanter to remove the sediment before drinking

Sieving

You may have used a sieve in the kitchen at home. A sieve is a special tool to help separate mixtures. The mixture often contains a liquid, usually water, and large pieces of something insoluble, such as vegetables or pasta. Scientists sometimes use sieves too. You may have used a stack of sieves to separate different sized pieces in soil samples.

Filtering

Sieving is only useful for removing large pieces from a mixture because sieves have quite large holes. They don't work if the material is in very tiny pieces. The small pieces of solid could pass through the holes in a sieve. If the material is insoluble and it is mixed with water the tiny pieces may float around in the liquid, giving it an opaque or milky appearance. We call this type of mixture a suspension.

◼ A sieve is useful for separating large pieces from a mixture

◼ Apparatus for filtering

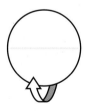
◼ Step 1: Fold in half.

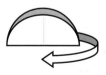
◼ Step 2: Fold into quarters.

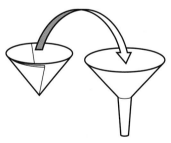
◼ Step 3: Open out to make a cone shape with three layers on one side and one layer on the other.

To separate a suspension, we need a special kind of sieve with tiny holes, so small we cannot see them. They need to be large enough for the water particles to go through but small enough to stop pieces of solid passing through. This kind of sieve is called a filter and is often made from a special type of paper called filter paper. To filter a mixture we fold the filter paper and put it into a funnel and then pour the mixture through it. The liquid will pass through the paper but the larger, insoluble material will get stuck in the paper. We call this the residue. The liquid that passes through the paper is called the filtrate.

Evaporating

Decanting, sieving and filtering will only work if you are separating out pieces of an insoluble material. When you are dealing with a soluble material you need a different process. A soluble material is one that dissolves when mixed with water. The crystals break up into such tiny particles that they can pass through the holes in filter paper, so they will remain mixed with the liquid in the filtrate. We cannot use filtering to separate a solute (dissolved substance) from a solution.

We know that evaporation is the process of a liquid turning into a gas or vapour. When you investigated the water samples earlier in this chapter you allowed the water to evaporate and so separated the mixture of water and dissolved material in the samples. When the water (the solvent) evaporated it became mixed with gases in the air but the dissolved materials in the water (the solutes) were left behind.

Evaporation is often used in science. Scientists quite often need to obtain dry samples of chemicals from solutions. You can do this by allowing the solvent to evaporate slowly, as you did with the water samples, but this can take a long time. We can speed up the process by heating the mixture gently. You can do this by placing it near a radiator, on a hot plate or, if you have the right apparatus, using the method shown in the picture.

It is very important to remember never to go on heating the solution until it is completely dry. If you do, the hot, solid material may start to spit out and could hurt you. You could also heat the material too much and this might change it in some way. You should always stop heating when there is still a little moisture in the container and allow the last little bit of solvent to evaporate by itself.

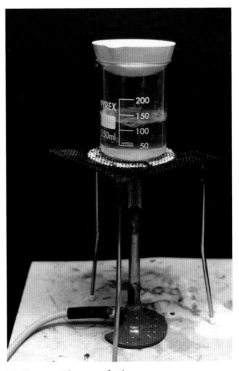
Evaporating a solution

Exercise 6.2a

1 What is the meaning of the term 'solvent'?

2 In a solution of salt in water, which of the substances is the solute?

3 What is a suspension?

4 Which separation method would be best for separating the following mixtures:

   (a) water and gravel

   (b) a solution of salt in water

   (c) a suspension of powdered chalk in water.

5 Explain why you could not separate the salts out of sea water by filtering.

6 When evaporating a solution, why should you not heat it until it is completely dry?

7 If you filtered a mixture of soil and water, which material would become the residue?

Exercise 6.2b

Joe and Laura are given a mixture by their teacher. The mixture contains the following materials: wax pellets, powdered charcoal, salt and iron filings. They are also given a sieve with holes 2 mm in diameter, some beakers, some water, a funnel and filter paper and a magnet.

1 Copy and complete the following table to show the properties of the four things in the mixture.

Material	Is it soluble?	Are the pieces larger than 2 mm?	Is it magnetic?	Does it float on water?
Wax pellets		yes		yes
Powdered charcoal	no	no		yes
Salt		no		
Iron filings		no		

2 Suggest a method that Joe and Laura could use in order to separate all four of the materials in their mixture.

The rock salt challenge

Rock salt is the material spread on icy roads in winter. It contains salt, which makes the ice melt more easily, and pieces of sand and grit, which increase friction so that cars do not slide around so much. Rock salt is a mixture.

Your teacher will give you a sample of rock salt. Your challenge is to separate the mixture so that you finish up with two piles: one of clean white salt and the other of sand and grit.

You will need to use more than one of the methods that you have learnt about in this chapter. You will also need to think about what you learnt in the last chapter about soluble and insoluble materials and making solutions.

You should plan your method and ask your teacher to check it before you start the experiment. Remember to think about safety. What precautions do you need to take in order to keep yourself and others safe?

Go further

⇨ How we can use separation

Common salt

Common salt, usually known simply as salt, is a very important chemical. It is an essential part of the diet of most animals, including humans, and for many centuries it was an important way of preserving foods, especially meat. Salt was a very valuable material. Countries that had a lot of salt used it to trade with countries that did not have a supply of their own. For example, the Ancient Egyptians traded salt for things such as timber and glass. Long caravans of camels carried salt across the Sahara desert right up until the late twentieth century.

Common salt can be obtained from two main sources. The first is the sea. In hot countries, sea water can be contained in shallow areas called lagoons or salt pans. The water then evaporates in the sun and the salt is left behind.

In cooler countries, such as Britain, there is not enough sunshine to make this method possible. Instead, sea water is filtered and then heated to evaporate the water. If salt solution is allowed to dry slowly, salt crystals begin to form in lovely cube or pyramid shapes. If a crystal of salt is allowed to grow for a long time it develops into a large cube.

The second source of salt is salt mines. Millions of years ago, some places that are now dry land were under the water of tropical oceans. As these oceans dried up, their salt was left behind in thick layers. These layers became covered in sediments of sand and silt to become an area of sedimentary rock. The rocks were then lifted by the movement of the Earth's crust and so the salt is now found in these rocks on dry land.

This rock salt is mined using huge grinding machines and is crushed for use on the roads and for other industrial processes. Some of it is also purified for eating, using a very large-scale version of the method that you will have used when separating your sample of rock salt.

■ Salt can be obtained by evaporating sea water

■ Salt forms cube-shaped crystals

■ In some places, salt can be mined from rocks

Oil chemicals

Oil is formed from the remains of tiny sea creatures. Millions of years ago, when they died and fell to the bottom of the sea, they became covered with layers of sand and silt. The heat and pressure of the sand and silt then changed the chemicals that made up the bodies of the animals into the chemicals that make up the mixture we call oil.

The oil we pump from the ground through oil wells is not very useful in its natural state. It contains hundreds of different chemicals mixed together. Some are very thick and sticky, such as the tar we use for road surfaces. Some are very light liquids that

evaporate easily, such as the petrol we put into cars, and some are gases. There are many different products that can be obtained from oil. These different products are separated using a process called 'fractional distillation'. This takes place at oil refineries. You will learn more about this process later. It involves heating the mixture very carefully. Different chemicals evaporate at different temperatures because they all have different boiling points. This makes it possible to separate them.

◼ In an oil refinery, oil can be separated to give useful products

Oil is important to us because it provides many products that we take for granted. There is oil for heating, diesel and petrol for cars and lorries and aviation fuel for aircraft. Chemicals from oil are used to make plastics, candles, clothing and paint, as well as lubricants to keep machinery working smoothly and for many other things as well.

Unfortunately, we are using up oil very quickly and the time will come when these products will no longer be readily available because oil will be so scarce. One of the big challenges for chemists is to find new ways of providing all these products using materials that will not run out. One example of this could be bio-fuels. These are fuels made from plant material that can be used in vehicles instead of petrol or diesel. This could be one answer to the problem. However, we will need to find ways of growing enough plant material to make this fuel without cutting down forests or taking up land that is needed to grow food. Products that can be made from materials that are replaceable and do not harm the environment or endanger our well-being are described as sustainable.

Chemists and materials scientists are very important people. Without them we would struggle to find new, sustainable ways of making the things that we need without using oil. Maybe you will be one of these important scientists when you grow up!

Exercise 6.3a

1 Explain why common salt was so valuable to people in past centuries.

2 (a) Describe how salt can be obtained from the sea in hot countries.

(b) How is sea salt obtained in Britain?

3 Salt mines contain layers of salt in between layers of rock. Explain how these layers of salt were formed.

4 Describe in your own words how oil is formed.

5 Name some products that we obtain from oil.

6 What name is given to the process used in oil refineries to separate the chemicals in oil?

7 Why do scientists need to look for alternatives to oil?

8 What is bio-fuel?

9 What is meant by the term 'sustainable'?

10 What do we have to do to make sure that bio-fuel is produced in a sustainable way?

Exercise 6.3b

Use the words below to complete the sentences. Each word may be used once, more than once or not at all.

biofuel   cook   glass   hundreds   meat   millions   mines

petrol   plastics   preserve   sea   sustainable   timber

1 Common salt was once used to _____ food, especially _____ .

2 The Egyptians traded salt for products such as _____ and _____ .

3 Common salt can be obtained from the _____ or from _____ .

4 Oil was formed from the bodies of tiny _____ creatures that died _____ of years ago.

5 Chemicals from oil are used to make products such as _____ and _____ .

6 Instead of petrol we can run cars on _____ which is more _____ than oil.

Drawing science diagrams

Working Scientifically

When scientists record the way they carry out an experiment, they often need to show the apparatus they have used. One way of doing this might be to take a photograph but sometimes photographs are not very clear and it is difficult to avoid getting all sorts of extra things in the background.

The clearest way is to draw a diagram. Diagrams are not pictures. We do not try to draw our apparatus in 3D because it is difficult to do so clearly unless you are a very good artist. Instead, we use a simplified set of shapes that anyone can draw to show the different pieces of apparatus.

You should always use a sharp pencil and a ruler when drawing these diagrams.

Your diagrams should not be too small, as it is hard to make tiny diagrams clear enough, nor too large. Your teacher will help you to get the size right. The really important thing is to make them neat and clear. Here are the shapes that we use for the apparatus you may have used in this chapter.

This is how we draw a beaker. Notice that there is no line across the top. The beaker also has a flat base, always drawn using a ruler, but has curved corners.

This is how we draw a funnel. Notice again that there is no line across the top and also that there is no line across the bottom of the spout.

To show the filter paper in the funnel, we draw a v-shape using a dashed line.

You may have supported your funnel in a conical flask. This is how you draw a conical flask.

There are some things that we do not bother to draw because they are not important. For instance, we need to show that our apparatus is resting on a flat surface. However, it really doesn't matter whether that flat surface is a table, a laboratory bench or the floor so we do not bother to draw these; we just draw a single straight line, using a ruler, under our diagram.

■ Remember to draw both the bottom of the beaker and the surface it is resting on

We do not draw clamps or filter rings either. All that matters is to show that the apparatus is supported in some way. It is not important what is supporting it. You could even be holding it yourself. We show the support with two neat crosses at the sides of the apparatus where the support is gripping it.

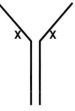

■ Supports such as clamps are shown by two small neat crosses

We do not draw heaters or Bunsen burners in diagrams. When we want to show that something is being heated we can do this by drawing a neat triangle or an arrow under it with the word 'HEAT' written neatly below it. Some people like to do this with a thin red pen or sharp red crayon – your teacher will tell you how you should do it.

The other thing that we do not draw is you! There is no science diagram shape for a person, so do not try to draw yourself in the diagram.

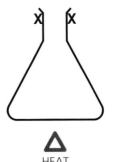

HEAT
■ A heater or burner can be shown by a small triangle

Diagrams should always be labelled. You do not need to label everything – just enough to make it clear. If you have drawn your diagram neatly and properly, it is probably not necessary to label the obvious things like a beaker. Labels should be written neatly at the side of the diagram to stop them from cluttering up your nice neat drawing. Then use a pencil and a ruler to draw a straight line from the label to the correct part of the diagram.

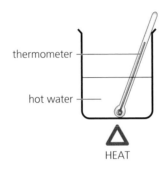

We can now put all these things together to show how the apparatus was used to filter a mixture:

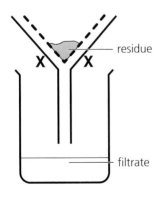 **or**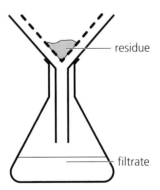

1 Make neat copies of the diagrams for the following apparatus:

(a) a beaker

(b) a funnel with filter paper in it

(c) a conical flask.

2 Now practise drawing the same diagrams without copying them.

3 Draw a neat, labelled diagram of an arrangement of apparatus that you could use to separate a mixture of chalk powder and water by filtering.

Go further

Chromatography

The term chromatography comes from the Greek words for colour and writing. Chromatography is a method used to separate a mixture of chemicals that are all dissolved and mixed together, such as the mixture of dyes used to make coloured inks. There are a number of different ways of carrying out chromatography. Some require expensive machines and some are quite simple. The method you will use is one of the simple ones and is called paper chromatography.

Any absorbent paper can be used for chromatography but filter paper or special chromatography paper works best.

Activity – separate coloured inks

You will need:

● a strip of filter paper or chromatography paper

● a pencil

● a coloured felt pen

● a test tube

● some water.

Draw a line in pencil across the paper, about 1.5 cm up from the bottom.

Use the felt pen to draw a circle of ink on the pencil line, about the size of a pea.

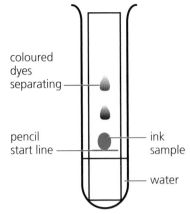

coloured dyes separating

pencil start line

ink sample

water

Put about 1 cm depth of water in the test tube. Check that the ink circle is above the water level when you put the paper strip into the test tube. This is important because, if the ink is submerged in the water, it will be washed away into the water and there will be none left on the paper for you to test.

Drop the strip of paper carefully into the test tube and watch what happens as the liquid soaks into the paper.

When the water level nearly reaches the top of the paper, take it out and let it dry. It is now called a chromatogram. Can you see which colours of ink were mixed together when making your felt pen?

When you put the paper strip into the water in the experiment above, you will have noticed that the water was quickly absorbed by the paper. The water level rose up the paper and carried the colours with it. Each colour travelled a different distance and so they became separated.

How does this work? When the bottom of the paper is submerged in the water, it becomes very wet. The water is absorbed and gradually moves up, higher and higher up the paper. The soluble inks dissolve in the water and are carried up the paper. Higher up the paper, the amount of water in the paper gets less and less. Gradually the coloured dyes get left behind. The more soluble the dye, the further it is carried up the paper. Any dye that is not soluble in water will not move at all. It will remain in the place where you drew the dot. You should now be able to look at your chromatogram and work out which dye was the most soluble and which was the least soluble.

The base line on a chromatogram must be drawn in pencil because, if you drew it in ink, the dyes making that ink would travel up the paper with the dyes from the felt pen dot and spoil your chromatogram.

Exercise 6.5

1 What is meant by the term 'chromatography'?

2 Give two examples where chromatography might be used to find out more about some mixtures.

3 What is a chromatogram?

4 Where would you find the most soluble dyes on a chromatogram?

5 If a dye did not move up the paper at all, what does this tell you about the dye?

6 Draw a neat, labelled diagram showing how you would set up a chromatography experiment.

7 Chemical changes

⇨ What is a chemical change?

You have already learnt about some of the changes that take place when materials are heated or cooled. For instance, you learnt that ice changes to liquid water when heated; we call this change melting. Water can be frozen again to make ice, so melting is a reversible change. We can change water from a solid to a liquid and back to a solid again as many times as we like. It is still water, even though it has a different name when it is a solid (ice). Nothing new is made when this change takes place. We change only the state of the water. We also learnt about evaporation and condensation. These changes are also reversible.

■ Some changes of state are reversible

Sometimes when we heat materials something different happens. The process of heating changes the material so that a new substance or material is formed.

Here are some everyday examples. If we put an egg into hot water or into a hot frying pan the egg changes. The translucent egg white becomes opaque and solid and the runny yolk becomes harder. We cannot change it back into the runny egg that came out of the shell.

When we make a cake, we mix flour, sugar, butter and egg together and put the mixture into the oven. When it comes out, the mixture looks quite different. We cannot change the cake back into the ingredients again. A non-reversible change has happened.

■ Cooking is a non-reversible change

In these examples something new is made when the material is heated. We call this type of change a chemical change or chemical reaction. The new materials are called the products of the reaction. Sometimes these new materials may be useful.

Chemical changes are happening all around us. Some are completely natural. For instance a chemical change takes place when a peach ripens and changes from a hard, sour fruit to one that is soft and sweet. A chemical change takes place in milk if it is left in a warm place and turns it sour. Other chemical changes are man-made. The cement used by a bricklayer between the bricks in a wall is wet and sticky. A chemical change takes place to turn it into the hard, strong material that holds the wall together. Without these chemical changes our lives would be very different.

Exercise 7.1

Are the following changes reversible or non-reversible?

1 melting chocolate

2 burning a candle

3 baking bread

4 freezing juice to make an ice lolly

5 drying clothes

⇨ An important chemical change – combustion

Combustion is the scientific name for burning. You will have seen things burning such as bonfires, candles or wood-burning stoves.

Sometimes things burn when we do not want them to. Fires in forests and woods can cause a lot of damage. Sometimes these fires reach people's homes and can kill people. Fire is dangerous, so it is important to treat it very carefully. You should never play around with fire as you could get hurt or cause a fire in your home, school or elsewhere.

■ Combustion is a dangerous non-reversible change

For combustion to take place three important things are needed. There has to be:

● a fuel, something that will burn (This could be a fossil fuel, such as solid coal, liquid oil or gas. It could also be wood, paper or candle wax. Many materials make good fuels.)

- an air supply (or oxygen)

- a source of heat to start the reaction.

These three things are often shown on a diagram called a fire triangle.

If one of these three factors is missing, combustion cannot occur. Fire fighters know this and so, when they put out a fire, they usually spray lots of water onto it. This removes the heat and also blocks the air from the fire. Some fires, however, cannot be put out using water. For example, you should never pour water onto a fire in a frying pan or chip fryer. The fire is very hot and the water would quickly turn to steam and make the burning fat spit out all over the place. This would make the fire worse. Can you explain why water should never be used on a fire in an electrical device?

Sometimes a fire blanket can be used to put out a fire. Can you explain how this would work? Have a look at the fire extinguishers around your school *but do not touch them.* You may be able to see a label telling you what they have in them. Some contain foam and some contain carbon dioxide. Both of these put out fires by blocking the air supply.

Activity – make a fire extinguisher

You should wear eye protection to do this experiment and should not do it without an adult to help you.

You will need:

- a plastic bag (with either a zip lock or wire tie)

- vinegar, 2 tablespoons

- bicarbonate of soda (sodium hydrogen carbonate), 1 tablespoon

- a candle securely held in a sand tray

- matches or a lighter

- eye protection.

Open your bag and put the bicarbonate of soda into the bag.

Pour the vinegar into the bag and then quickly seal the bag with the wire tie or the zip lock.

Watch what happens to the bag.

When the vinegar and the bicarbonate of soda come together a non-reversible chemical reaction takes place. New products are formed. One of the products is a gas, carbon dioxide. Because the gas takes up more space than the materials you put into the bag, the bag inflates like a balloon.

Ask an adult to light the candle for you. Ask the adult to help you hold the bag near the candle but *not directly over the flame*. Position the bag so that the opening points towards the flame and open it carefully. Carefully tip the opening of the bag slightly towards the candle, making sure that none of the liquid comes out. The gas will pour out over the candle and put it out. Pure carbon dioxide is heavier than the mixture of gases that make up the air so it sinks down and surrounds the candle, stopping the air from reaching it.

Did you know?

Bicarbonate of soda is used in baking to make cakes rise. It reacts with other materials in the cake mix and makes bubbles of carbon dioxide. The bubbles get trapped in the cake mix and the cake rises.

Exercise 7.2a

Use the words below to complete the following sentences. Each word may be used once, more than once or not at all.

air	carbon dioxide	coal	combustion
electrical	extinguisher	fat	fuel
gas	heat	heavier	lighter
oil	oxygen	water	

1 The scientific name for burning is _____.

2 The three parts of the fire triangle are _____ , _____ and _____.

3 The three types of fossil fuel are _____ , _____ and _____.

4 A fire _____ puts out a fire by blocking the _____ supply.

5 Water should never be used to put out fires in _____ devices or burning _____.

6 _____ _____ gas is used to put out fires because it is _____ than air.

Exercise 7.2b: extension

Make a fire safety leaflet or poster designed for other children of your age. Make sure that your work is clear and that you explain your message carefully.

More about candles

Have you ever wondered what happens when a candle burns? We all know that the candle gets smaller when it burns but why does this happen? Where does the candle wax go?

Let's see if we can work out what is happening.

■ What really happens when a candle burns?

Working Scientifically

Activity – investigate a burning candle

You should wear eye protection to do this experiment and should not do it without an adult to help you.

Getting smaller?
You will need:

● a candle

● scales

● a sand tray and/or a candle holder

- a match or lighter

- eye protection.

Weigh the candle accurately and record its mass.

Fix the candle securely in a sand tray or candle holder. Put it in a place where it will not be knocked over or brushed against. Ask an adult to check that it is safe.

Now ask an adult to help you to light the candle safely and leave it to burn for about half an hour.

Blow out the candle very carefully. Make sure you do not blow away any melted wax from the candle. Leave it to cool.

Predict what you will find when you weigh the candle again. How heavy do you think it will be? Remember to explain why.

When the candle is cool and the wax has all solidified, weigh it again. Was your prediction correct?

The candle gets lighter when it burns. It gets shorter too and we cannot bring the candle back to its original size. A non-reversible chemical reaction has taken place. The wax has been changed into new substances. These substances seem to disappear because they are gases and they mix with the air.

Remember the fire triangle. Where did heat to start the combustion reaction come from? What is the fuel?

Often people think that the wick is the fuel. The wick is usually made from cotton string. It actually does not burn very well but glows hot enough to melt the candle wax and then evaporate it. It is the wax vapour that burns.

The air supply
The fire triangle tells us that an air supply is needed for things to burn. In these experiments you will find out more about this.

You will need:

- a night-light candle

- a large bowl of water

- a large jam jar

- eye protection.

Float the candle in the bowl of water.

Ask an adult to help you to light the candle safely.

Turn the jam jar upside down and carefully lower it over the candle so that the opening of the jar is just under the surface of the water. Hold it steady and watch very carefully what happens. Can you suggest an explanation for your observations?

At first you might have noticed that the air in the jar pushed the water level down a little when the jar was first put into the water. After a short time, the candle goes out and then the water level rises up inside the jar. What causes this to happen?

Here is another experiment to try.

You will need:

- a night-light candle
- a sand tray
- a match or lighter
- jam jars of different sizes
- a timer
- eye protection.

Place the candle safely in the sand tray.

Ask an adult to help you to light the candle safely.

Carefully place one of the jam jars upside down over the candle and push it down into the sand.

Watch what happens.

Try this again with the same jam jar and time how long the candle takes to go out.

Now try with the other jam jars. Predict how long you think it will take for each candle to go out. Can you see a pattern in the results? What could you do to make sure that your results are as reliable as possible?

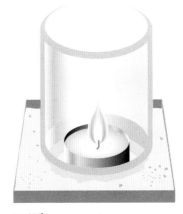

When you put a cover over the candle like this, the candle eventually goes out. Why do you think this happens?

The candle needs oxygen from the air for the reaction we call combustion. It can only burn when there is enough oxygen (at least 15 per cent) in the air. Discuss your observations with your partner or group. Can you use this fact to help you to explain your observations from these two experiments?

In the experiment with the floating candle you saw the water level rise inside the jar. The candle used up some of the air as it burned. This means that there was less air and so the water rose in the jar to take the place of the air that had been used up.

When you put jars of different sizes over the candle, the candle went out. The larger the jar, the longer the candle should have burnt before it went out. The candle needs something from the air to burn.

About four-fifths of air is a gas called nitrogen. Most of the rest of the air is oxygen and this is what the candle needs. When the candle burned it used the oxygen in the jar. The candle went out when there was not enough oxygen left to keep it burning.

The products of the reaction

Combustion is a chemical reaction and all chemical reactions produce new substances. In this experiment you will find out a bit about the new substances that are formed when a candle burns.

You will need:

- a candle
- a sand tray and/or candle holder
- a match or lighter
- a large jam jar
- eye protection.

Place the candle carefully in the sand tray or candle holder and ask an adult to check that it is safe.

Ask an adult to help you to light the candle safely.

Turn the jam jar over and place it over the candle as before and watch carefully.

■ When a candle burns, the products of the reaction are gases

You already know what will happen to the candle. This time look carefully at the top and sides of the jam jar. What can you see?

When the candle burned in the jar, did you see the jar becoming misty as water vapour condensed on the sides of the jar? Water vapour is one of the products of the combustion reaction.

Another product created is the gas carbon dioxide. You may remember that you learnt in Chapter 4 that this gas is formed when fuels burn.

We can test whether a gas is carbon dioxide by using a liquid called limewater. This looks like ordinary, clear water but when it is mixed with carbon dioxide, a reaction occurs and a white chalky suspension is formed, making the limewater go cloudy. Your teacher may help you to carry out an experiment to test for carbon dioxide when a candle burns.

If a candle is put into a large jam jar, lit and allowed to burn for a while and then carefully extinguished, we can pour a little limewater into the bottom of the jar and shake it around a bit. The limewater goes cloudy because the carbon dioxide formed when the candle was burning has stayed in the jar.

■ Limewater is used to test for carbon dioxide

When combustion takes place, the candle gets smaller because the particles making up the wax are used up when it burns, forming the gases carbon dioxide and water vapour. These gas products are then lost into the air. You may also have noticed some soot forming in the jam jar when you put it over the candle. Soot is formed during combustion if some of the fuel does not react fully with oxygen.

■ Limewater goes milky

Scientists often describe chemical reactions using something called a word equation. The word equation for combustion looks like this:

fuel + oxygen $\xrightarrow{\text{react to form}}$ water + carbon dioxide

Exercise 7.3a

1 Why does a candle go out if it is covered with a jar?

2 Name two gases that are found in the air.

3 Which gas makes up over three-quarters of the air?

4 Which gas in the air is needed for combustion?

5 What products are formed when a fuel such as candle wax burns in air?

6 Which of the products in Question 5 can be identified using limewater?

Exercise 7.3b

Here are some statements about burning a candle. Some are correct and some are incorrect. For each one say if it is true or false. Rewrite the false statements, correcting them to make them true.

1 A candle needs oxygen to burn.

2 A candle in a jar goes out when it has used up all the oxygen.

3 In a burning candle the fuel is the wick.

4 Most of the air is made up of nitrogen.

5 The new substances made when a candle burns are gases.

6 Lime juice is used to test for carbon dioxide.

Exercise 7.3c: extension

Peter thinks that when a candle burns, all the wax melts and drips down the side. He says that the wax could be collected and made into a candle again, so combustion is a reversible change. What could you say and do to show Peter that he is wrong and that combustion is a non-reversible reaction?

⇨ Another chemical change – rusting

The car in this picture is very rusty. You must have seen other things go rusty too. How do you think things become rusty? Do all materials go rusty? Discuss your ideas with your partner or group.

You have probably come up with several ideas. The children in this picture have also had some ideas. What do you think of their ideas?

To test your ideas you will need to do some experiments. First we will set up some experiments to find out whether it

■ Rusting is a chemical change

is water, air or both that make nails rust. You may be able to set up some extra experiments to test some of your other ideas as well.

I think that there is rust inside the nail. The rain washes away the surface and exposes the rust.

I think the rust is a layer of microbes in the air which settle on the nail.

I think that rust is made when metals and water react together.

I think that iron in the metal reacts with oxygen from the air.

Activity – what causes rusting?

You will need:

- iron nails
- test tubes or other containers
- rubber stoppers or lids to seal your containers
- boiled water
- oil
- silica gel
- labels.

In order to find out which conditions are necessary for rusting, we need to set up situations where we can take away one condition at a time. By doing this we can see which conditions are needed for rust to appear and which are not. We need to keep all the other conditions the same to make our tests fair. If the nail does not rust then the condition we took away is probably needed for rusting to occur. So that we can be sure that this conclusion is valid we need to have a control. Can you remember what a control is?

First set up your control experiment. Put a nail in a clean, dry container with no lid. Leave this out in the room during the experiment. You can check whether it has rusted and use it to compare with your other results. Label the container CONTROL.

Now set up the other conditions:

(a) nail in contact with water and air

Place the nail in a container with a little water or maybe with some wet cotton wool or tissue. Put a lid on the container to keep the moisture in. There will be plenty of air trapped in the container. Label the container WATER AND AIR.

(b) nail in contact with water but no air

Water contains dissolved air so the water must be boiled for a few minutes to remove the dissolved air. Your teacher will do this for you. Place your nail in the container and cover it with the boiled water. You can then float a layer of oil on the surface to act as a seal to keep the air out and stop any more air from dissolving in the water again. Label the container WATER, NO AIR.

(c) nail in contact with air but no water

You need to seal the nail in a completely dry container. There is water vapour in the air so this needs to be removed. We can do this by putting a substance that absorbs water, usually either silica gel or anhydrous calcium chloride, inside the dry container. It will absorb the water vapour and dry the air. You then need to seal the container tightly to make sure that no water vapour from the outside air gets in. Label the container AIR, NO WATER.

Set up any other experiments that you and your class want to try. Remember to label the containers to show what conditions the nails are in.

Place all your containers together somewhere where they can be left for a few days.

Which nails do you think will rust? Write down your predictions so you can remember them when you look at the results in a few days' time.

After a few days look at your nails carefully and record how much rust you can see on each one. What conclusion can you draw from these results about what conditions cause iron to rust?

Your experiments should have shown that air and water together cause objects containing iron to rust. The damp iron reacts with oxygen from the air and makes a new substance, which we call rust. Did you find out anything else interesting about rusting in your experiments?

Iron is the only metal that rusts. Many objects you see becoming rusty are made from steel. Steel is a mixture of substances, including iron and carbon. It is the iron in steel that rusts. Other metals do react with oxygen but they do not form the reddish brown material we call rust. Silver tarnishes to make a dull grey coating. When copper is left in the air, it often forms a green substance called verdigris. Some metals react so quickly with oxygen that they burst into flames! We have to be very careful to store these so that no air can get to them.

■ Copper turns green when it reacts with the air

In the combustion experiments we saw that air is used up when fuels burn. Does rusting use up air as well?

We have shown that rusting takes place when iron or steel comes into contact with air and water. We suggested that the iron reacted with the oxygen from the air. We can do an experiment to show that rusting uses up air.

Activity – using up air

You will need:

- iron wool
- a large test tube (boiling tube)
- a bowl of water
- a retort stand and clamp or other way of supporting the test tube to stop it falling over.

Take some of the iron wool and push it to the bottom of the test tube. Do not pack it too tightly. Check it will not fall out when the test tube is turned upside down.

Put some water in the test tube and shake it about a bit to make sure that the iron wool is really wet. Tip out the remaining water.

Turn the test tube upside down and put the opening just into the water in the bowl. You need to make sure that the opening of the test tube is under the surface of the water but do not push it too far down or you will not be able to see the results.

Clamp the tube firmly and put it somewhere where it will not be knocked or touched for about a week.

Take a look at your experiment each day. Can you explain what happens?

When there is no further change taking place, use a waterproof pen to mark the level of the water in the test tube. How much air has the rusting iron wool used up?

When the iron wool rusts, it uses up air, just like the candle did when it burned. As the air is used up, water is drawn up into the test tube to take the place of the used air. The water level rises but when about one-fifth of the air has been used up, the level stops rising. Why is this?

You may remember that we learnt that the iron uses oxygen when it rusts. When all the oxygen has been used up, no more rusting can occur. The experiment shows that oxygen makes up about one-fifth of the air.

Preventing rusting

Generally, rusting is rather a nuisance. When iron or steel rusts, a strong metal structure can change to crumbly brown rust. Objects can be spoilt by rust but there are ways to stop it.

Since rusting is caused by a combination of water and oxygen coming into contact with the iron, the best way to prevent it is to make sure that water and oxygen are kept away from the surface of the object. There are lots of ways of doing this. For example, the surface can be covered in oil or paint. See how many other different ways you and your partner or group can think of to keep oxygen and water away from the iron. Perhaps you could do some experiments to try out some of your ideas.

■ Many structures are painted to protect them from rusting

Another way of keeping iron or steel objects from rusting is to cover them with a very thin layer of another metal that does not rust. This process is called galvanisation and the metal that is used is usually zinc.

First, the object is cleaned thoroughly, often by dipping it into a tank of hot acid. This removes dirt and any traces of rust from the surface of the object. It is then dipped into a tank of melted zinc. When it is taken out, the zinc sticks to the surface in a very thin layer. If you look closely at a galvanised surface, you can see beautiful crystals of zinc. This layer will stop the water and oxygen from reaching the surface of the iron or steel.

■ Look closely at the surface of a galvanised object, such as a watering can, to see the zinc crystals on the surface

Did you know?

When we buy soup or tomatoes from the supermarket, they may be stored in metal cans. People often call these 'tins' because the early cans were made from tinplate steel. The steel was coated with a very thin layer of a metal called tin to prevent rusting. Nowadays so-called 'tin cans' often have no tin in them at all. Some have a plastic coating and some are made from aluminium, which doesn't rust.

1 What conditions cause iron to rust?

2 How can dissolved air be removed from water?

3 When you have removed the air from some water, how can you stop more air dissolving in the water?

4 Name a substance that can be used to absorb water vapour from the air.

5 Explain why steel objects rust.

6 If a car rusts a lot, it must be treated and mended before its owner is allowed to drive it again. Explain why rusting might make a car unsafe.

7 Bicycle frames are usually made from steel and they are painted to prevent rusting. The chain of the bicycle is also steel but it is not painted.

(a) Why would painting the chain not be the best way to prevent rusting?

(b) Suggest how the chain could be kept rust free.

8 (a) What is meant by the term galvanisation?

(b) Explain in your own words how an object is galvanised.

Exercise 7.4b

1 Name two substances that are needed for iron to rust.

2 Why might a rusty car be unsafe?

3 To stop iron and steel rusting they can be covered in a thin layer of another metal.

(a) Which metal is usually used?

(b) What name is given to this process?

(c) How does this stop the iron or steel from rusting?

4 Give two other ways that iron and steel can be protected from rust.

Exercise 7.4c: extension

A sculptor makes a beautiful sculpture using steel. He likes the look of the shiny steel but he knows that it will soon rust when he puts it outside.

Write a letter to the sculptor, suggesting some ways in which he might be able to keep the sculpture looking shiny and new. Describe how he could carry out a series of experiments to test different methods of rust prevention, to see how well they work.

■ Sometimes artists want their work to rust; the rusty red colour of this fox sculpture makes it look almost real

➾ Natural or man-made?

Many of the materials that we use can be found around us. Wood and stone are good examples. Trees grow all over the place and stone can be dug out from the ground. We call these natural materials.

Some natural materials must be changed in some way before they are used. Clay is a natural material but it is not much use in its natural state. It is soft and floppy when it is wet. It is weak and crumbly

■ Stone is often used for building

when it is dry. If we make clay very hot in a special oven called a kiln, it changes so that it becomes hard and strong. Bricks, cups and wall tiles are all made from clay.

Glass is not a natural material. It is made by mixing together sand, ash and limestone, and heating them up to about 1700°C. Glass is an example of a man-made material. The hot glass is soft, and bottles and glasses are sometimes made by blowing a bubble into the middle of a lump of hot glass.

Another important group of man-made materials is plastics. Plastics are usually made from oil but they can be made from other things as well. Plastics are very useful. They are light and colourful. They are waterproof. They can be easily shaped too, so they are used to make lots of things. How many things can you see around you that are made from plastic?

One problem with plastic is that it is difficult to get rid of when we have finished with it. It does not rot away in rubbish tips and it can be harmful to wildlife.

■ Bricks are made from clay that has been heated to about 1000 °C

■ Glass blowing

⇨ Using chemical changes

Understanding the way that materials react when we mix them together allows scientists to explore ways of making new materials for industry. There are many people working hard to find new, environmentally friendly fuels and new ways of making materials such as plastics. The more we understand, the more new and exciting materials are being invented.

You may think that making new materials is hard but you can make some potentially useful new materials yourself. Here is an example.

Activity – making plastic

As we have already learnt, plastics are normally made from oil. However it is possible to make plastic from a much more sustainable raw material – milk.

You will need an adult to help you with this.

You will need:

- skimmed milk
- a measuring jug
- a saucepan
- a cooker or electric ring
- a tablespoon
- a sieve
- vinegar
- a biscuit cutter.

Measure out 300 cm³ of skimmed milk into a saucepan.

Ask an adult to help you warm the milk gently but do not let it boil.

Next, add a tablespoonful of vinegar and stir it until it goes lumpy.

Hold a sieve over the sink and carefully pour the mixture through to collect the solid lumps.

When the lumps are cool enough to handle squeeze the solid pieces together to make a ball.

Press your plastic firmly into a biscuit cutter to shape it and push it out gently. Leave it for a few days to harden.

Making better plastic
You have made some basic plastic but is it the best you could make? Discuss with your partner or group what changes you could make to this method and say what difference you think they might make. When you are ready, explain your ideas to your teacher and ask if you can try them out. Remember to change one thing at a time so you can tell what is making the difference.

Working Scientifically

Plastic in the ocean

Some of our waste plastic finishes up in the sea where it causes serious problems to animals that mistake the plastic for food. It has been estimated that up to 90 per cent of sea birds have plastic in their stomachs. Sea turtles often mistake floating plastic bags for jellyfish, their favourite food. Jellyfish are slippery and hard to eat so turtles have a special adaptation in their throat that stops slippery things from sliding back out. This means that, once a turtle has grabbed a plastic bag in its mouth it cannot usually spit it out again and has to swallow it.

One way to deal with this problem is to recycle plastic objects. Another is to find ways of using plastic boxes, bottles and bags again. Other materials can be recycled too. Do you have a recycling bank near you? Maybe you have special boxes in which to collect your recycling.

The oil from which plastic is made is a natural resource and we use a lot of it. Other useful things like petrol are made from oil. We will run out of oil one day, maybe in your lifetime. It is important that we try not to use more of it than we need.

Objects made from materials that cannot easily be recycled are usually taken to landfill sites. These are large pits, usually old quarries, where the waste is dumped into the ground and then covered over. You might think that this is the end of the problem but the rubbish in these pits begins to decay and release poisonous gases and chemicals and so they need to be carefully managed for a long time. We are also running out of places to put these landfill pits so people are working on better ways of getting rid of rubbish. One way is to burn it and use the heat to generate electricity. This has many advantages but it can also produce poisonous gases, which need to be carefully removed before the smoke is released into the air.

The best way for us to protect the environment is to have as little waste as possible.

■ Some materials do not rot away so it is important that they are used again or recycled

■ Landfill sites take up a lot of space and pollute the environment

Did you know?

Compost heaps are also a kind of recycling. All the garden and kitchen waste that is put in to the heap is broken down by decomposers such as bacteria, fungi, worms and woodlice. This forms humus which can be returned to the garden to provide nutrients for the growing plants and help to retain water in the soil.

■ Compost heaps recycle nutrients in the garden

Activity – using and recycling materials

1 Look at the materials that have been used to make the buildings around your school. Make a list with two columns. In one column, write all the natural materials and in the other, write the man-made materials.

2 Think about recycling in your school. Do you collect waste paper in your classrooms? Find out what other materials could be recycled and start a collection.

3 Have a recycling competition. Try to think of as many ways as possible to reuse a plastic water or soft drinks bottle. See whose idea is the most imaginative.

Exercise 7.5a

1 What is meant by the term 'natural material'?

2 Give two examples of natural materials.

3 What is meant by the term 'man-made material'?

4 Give two examples of man-made materials.

5 Which materials are needed to make glass?

6 List five things that are made from glass.

7 What is plastic usually made from?

8 Why do we need to try to avoid using too much plastic?

9 Explain why waste plastic can be a particular problem in the ocean.

10 Why should we try to reduce the amount of our rubbish that goes into landfill?

Exercise 7.5b

Use the words below to complete the following sentences. Each word may be used once, more than once or not at all.

clay	glass	food	landfill	man-made
natural	oil	plastic	recycling	sea
useful	wood			

1 Materials such as wood and rock that are found all around us are called _____ materials.

2 Some natural materials, such as _____, need to change before they become really _____.

3 Materials such as _____, which are made by putting two or more materials together, are called _____ materials.

4 _____ is a useful man-made material made from oil.

5 Waste _____ is a problem in the _____ because animals think that it is _____ and eat it.

6 Some materials will run out if we use too much of them. We can help to make them last longer by _____ them.

7 Rubbish that is not recycled is put into _____ sites.

Make an information leaflet or poster about recycling. It should be written for children of your own age and should explain why it is important to recycle materials. Include information about how to recycle or reuse materials in school and at home. Make it colourful and attractive. You could even include a quiz or competition to make it more fun.

⇨ People in science: making new materials

Spencer Silver and sticky notes

We have seen in this chapter that non-reversible reactions can result in the formation of new materials and that these new materials may be useful. Sometimes these useful materials are created by scientists by mistake. Spencer Silver worked for a company as a research scientist, trying to find a really strong adhesive (glue) for the space programme.

One day he made a mixture that was quite the opposite. This adhesive was really weak, so things stuck together with it could be peeled apart again really easily. It obviously was not the answer to Spencer Silver's research project. He was sure that it might be useful for something, but no one could quite work out what, so he went back to making strong adhesives. He never gave up on his discovery though and continued to try to think of ways of using it. One suggestion was to paint it onto notice boards so that pieces of paper could be put on and then taken off easily. However, this idea was not going to make enough money to make it worthwhile to produce.

For many years the strange glue was just an oddity. Then one of Spencer Silver's colleagues came up with an idea. He was a musician and needed a quick and easy

■ Sticky notes have many uses, even making pieces of art

way to mark places in his music. He suggested spraying the glue onto paper strips. The strips could be used to make markers and then be peeled off again without damaging the music. It was a good idea but at first it didn't work. The glue would not stay on the paper strips, but after some further research work this problem was solved and sticky notes were born. An experimental failure had been turned into a profitable product, thanks to the determination of Spencer Silver and his colleagues.

Activity – other discoveries

Spencer Silver's discovery has proved useful to many people. There are many other materials that we take for granted but most of them have been discovered by skilled scientists through careful research.

Here are some scientists who discovered new materials. Use books or the internet to find out what materials they discovered and what they are used for. You could extend this task by finding out more about their lives and work and writing a mini-biography of them.

- Ruth Berenito

- Wallace Carothers

- Stephanie Kwolek

- Charles Goodyear

- Roy Plunkett

8 Earth and space

⇨ A brief history

When we look up into a cloudless night sky, we can often see a beautiful collection of bright stars. There are many millions of stars in the universe, all grouped together into clusters called galaxies. The stars we can see in the night sky are the ones in our own galaxy, the Milky Way. The Milky Way is just one of millions of galaxies in the universe.

When we look up into the sky, it is hard to understand how far away everything is, but the distances in space are huge. The nearest star outside the solar system is called Proxima Centauri. It is trillions of kilometres away from us.

Did you know?

The distances in space are so enormous that scientists do not measure them in kilometres. They use a unit called a light year. One light year is the distance travelled by light, travelling at 299 792 458 metres per second, in one year. That is a huge distance — just over 9 trillion kilometres. Proxima Centauri is 4.22 light years away from Earth.

All over the world, for thousands of years, people have looked up into the sky and tried to make sense of what they could see. In many places, ancient people worshipped things they could see in the sky. Some believed that the Sun was a god, some worshipped the Moon. People often saw patterns in the stars and used these to forecast the future.

■ The Inca people of Peru worshipped the Sun

We still pick out patterns in the stars to help us to find our way around the night sky. These patterns are called constellations. Maybe you can recognise some. The best known are the Great Bear (sometimes called the saucepan or big dipper) and Orion the hunter.

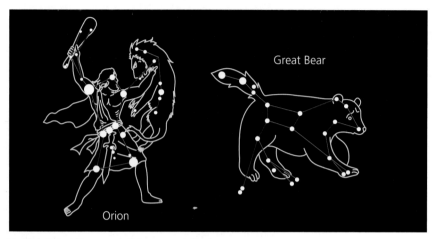

Great Bear

Orion

■ The Great Bear and Orion constellations

Activity – constellations

Use books or the internet to find the patterns of stars that make up some of the constellations. You could start by finding out about Cassiopeia, Draco and the Southern Cross. You could make the patterns by sticking silver stars onto black paper for a display.

Look at a newspaper, the internet or a magazine to find out about the night sky you could see tonight. Where will you look to find the North Star? Which constellations will be visible? Will you be able to see any of the planets?

People who study the universe are called astronomers. Early astronomers had to rely on what they could see with their naked eyes. Until the sixteenth century, people believed that the Earth was at the centre of the universe and that everything else moved around it. This belief was largely based on the work of a Greek man called Ptolomy, who lived in the 2nd century AD. He used his observations of the movement of the planets and some complicated mathematics to explain how the planets and the Sun moved round the Earth. His explanation was later challenged by an Arab scientist called Alhazen (965–1040). Alhazen pointed out that Ptolomy's observations and his mathematics did not match up, but his new approach still had the Earth at the centre of the Universe.

In 1530, an astronomer called Copernicus published a book explaining that the Earth was one of a number of planets that orbit (move around) the Sun. This idea was not completely new because a number of thinkers had suggested it earlier, but the idea had never been fully accepted. Another scientist, Galileo (1564–1642), also believed that the Earth moved round the Sun. He got into a lot of trouble with the leaders of the Roman Catholic Church who believed what the Bible said — that the Earth was at the centre of the universe.

Galileo was probably the first person to observe the stars and planets using a telescope. He made his own telescope, which was not at all powerful by today's standards, but it let him see the planets in more detail. He was the first person to identify some of Jupiter's moons.

Galileo made his own telescopes to observe the stars and planets

Nowadays we have telescopes that are much more powerful. Some are massive; some are out in space, orbiting the Earth. They allow us to see out into space, beyond the Milky Way, and to find out about distant stars and planets. Spacecraft have sent back pictures of the planets and their moons. Some people have even left the Earth and travelled to the Moon. All these things have helped us to find out more about the universe and especially our nearest neighbours in the solar system.

Did you know?

In 2013 and 2014 there were several 'firsts' in the world of space exploration. In 2013 the Chinese landed their first lunar spacecraft on the surface of the Moon.

In 2014 the Indian Space Research Agency's Mars Orbiter Mission successfully went into orbit around the red planet. This is India's first interplanetary mission.

Perhaps the most extraordinary 'first' in 2014 was the first spacecraft to go into orbit around a comet. The Rosetta mission was launched by the European Space Agency in 2004. During its ten year trek through space it crossed the asteroid belt and travelled millions of miles, more than five times the distance of the Earth from the Sun, to meet up with the comet (called Comet 67P/Churyumov-Gerasimenko). All this time, the instruments on board were placed in 'sleep mode' to save energy. They were successfully woken up again just before Rosetta reached the comet. The space probe went into orbit around the comet and then released a small lander, called Philae, which dropped to the surface of the comet – very slowly because of the very low gravitational field. Both Rosetta and Philae sent back amazing pictures and took lots of measurements which have helped us to understand more about comets.

We can see the Sun and other stars because they are luminous. Planets and moons can only be seen because they reflect light from the Sun from their surfaces into our eyes. This makes them much harder to spot than the stars. Some modern telescopes do not use visible light to study the universe. They look for other types of rays, such as infra-red and X-rays. This technology has allowed us to find out much more than would be possible using only visible light.

■ People first landed on the Moon in 1969

Activity – space missions

Find out about some of the spacecraft that have been launched from Earth and where they have visited. See if you can find any of the pictures they have sent back to Earth. You might like to start with Cassini, the Venus Express, some of the Apollo missions or the Hubble telescope.

Exercise 8.1

Use the words below to fill in the gaps in the sentences below. Each word may be used once, more than once or not at all.

astronomer   constellations   galaxies   Galileo   stars

Milky Way   Moon   planets   spacecraft   telescopes

1 The universe contains millions of _____ grouped together into clusters called _____.

2 Our galaxy is called the _____.

3 People often look for patterns in the stars. We call these patterns _____.

4 A scientist who studies the universe is called an _____.

5 Early astronomers could not see many of the planets because they did not have _____ to look through.

6 _____ was the first astronomer to identify some of the moons of Jupiter.

7 The furthest that any person has travelled from Earth is to the _____.

⇨ The solar system

The Sun is a star at the centre of our solar system. It is a huge ball of burning gas and is incredibly hot. It is about 15 million °C at the centre. The Sun is about 150 million kilometres away from the Earth. The heat from the Sun that reaches the Earth makes the Earth a good place to live. Remember it is not safe to look directly at the Sun – even if you are wearing sunglasses.

Orbiting the Sun are the planets. There are eight planets in our solar system, all named after gods and people from mythology, apart from the Earth. The planets are all roughly spherical (ball-shaped), some are made from rocks and some are made from gas. They all travel around the Sun because the Sun's gravity pulls on them and stops them escaping into space. You will learn more about gravity at the end of this chapter. If we could travel from the Sun into the solar system, visiting the planets in turn, what would we see?

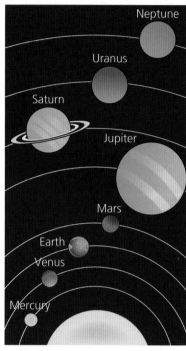

The solar system

The planet nearest the Sun is called Mercury. The Romans believed that Mercury was the messenger to their gods. Mercury is a rocky planet that gets amazingly hot in the daytime (about 350 °C) and extremely cold at night (−178 °C).

After Mercury comes Venus. Venus was the Roman goddess of beauty, but the planet Venus would not be a beautiful place to live. It is very hot and has thick acidic clouds and very strong winds. One spacecraft landed on Venus but only managed to send messages back to Earth for less than an hour before its instruments were destroyed by the heat.

The third planet from the Sun is Earth. As far as we know, Earth is the only place in the solar system where life exists. This is because living things need water. Earth is the only place where the temperature is not so hot that water evaporates away or so cold that it is all frozen solid. Maybe we will find life somewhere else, either in our solar system or on a planet orbiting another star somewhere in the universe. Scientists have found many planets orbiting other stars in the Milky Way and some of them could possibly have the right conditions to support life.

Planet number four is Mars, named after the Roman god of war. Mars is sometimes called the Red Planet because its surface is covered with red dust and rocks. We know a lot about Mars because many space probes have visited it and taken pictures of it.

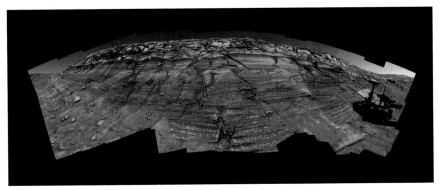

■ Rovers send back lots of information about the surface of Mars

In 2003, two vehicles or 'rovers', called Spirit and Opportunity, landed on Mars. Another rover, Curiosity, landed on Mars in 2012. They have been driving around, taking pictures and samples of rock and soil to help scientists to find out much more about this planet. One of the things scientists have been trying to find is evidence that there might once have been life on Mars. We now believe there was once liquid water on the surface because we can see features that look like dry rivers. There might have been living things on the planet, and it is not impossible that there is still life there, but they would probably have been tiny things like bacteria. There certainly have never been little green men!

If people ever travel farther away from the Earth, Mars is the most likely place for them to go. It would take more than five months to get there and life would be quite difficult. The lack of water would be a problem and there is also no oxygen in the Martian atmosphere. As it is further from the Sun than Earth, it can be very cold, maybe as low as −140 °C at the poles. Even the hottest day is below 20 °C, which is a comfortable room temperature on Earth. People travelling to Mars would also be exposed to a lot of harmful radiation from the Sun because Mars does not have a protective atmosphere like Earth.

Next on our journey would be Jupiter. The Romans named Jupiter after the king of their gods. Jupiter is the biggest of the planets. It is a huge ball of gas with a tiny rocky centre. There are huge storms on Jupiter. One of them is a massive swirling thunderstorm, about 25 000 km across, that may have been raging for over 300 years! This can be seen from Earth and is called the Great Red Spot. Jupiter has four large moons and 59 smaller ones that we know about.

Saturn is the most easily recognisable planet because of its beautiful icy rings. Saturn is also a gas planet and is named after Jupiter's father, who was the god of agriculture. In 1997, a space probe called Cassini was launched from Earth to go to Saturn. It arrived in 2004, a journey of seven years! Cassini has sent back some beautiful pictures of Saturn and its moons.

■ This picture of Saturn was taken by the Cassini spacecraft

Did you know?

Saturn is so light that it would float on water – if you could find a lake big enough to put it in!

The last two planets are called Uranus and Neptune. Uranus was the Roman god of the heavens and Neptune was their god of the sea. Both of these planets are spheres of gas and look blue. We do not know much about these planets because they are so far away. Neptune has really dreadful weather, with winds blowing up to 2500 km per hour!

Beyond Neptune comes Pluto. Pluto used to be included as one of the planets, but in 2006 astronomers decided that it was too small to be counted as a proper planet so it is now called a 'dwarf planet'. There are a number of other rocky objects orbiting in the area just beyond Pluto and some of these are as big or even bigger than Pluto. They are so far away that we have not yet been able to study them properly.

Activity – planets

1 Make a planets fact book. You could work as teams in your class to make a planet page each, or maybe do this on your own. Find out some interesting facts about each of the planets. For example, how far from the Sun they are, how long their days and years are, what they are made of, how many moons they have, how hot and cold they become. You could find a picture of each planet to illustrate your pages.

2 Find out more about what it might be like to land on Venus. Imagine that you are travelling to Venus and write a diary about your experiences. You could write about how you feel as you travel and what it feels like to pass through the clouds and land on the planet.

Exercise 8.2a

1 What is the Sun?

2 How far away from Earth is the Sun?

3 How long does it take light to travel from the Sun to Earth?

4 How many planets are there in the solar system?

5 Which planets are closer to the Sun than the Earth?

6 Which planet is named after the Roman god of war?

7 How do we know so much about the planet you named in Question 6?

8 Why is Earth the best place in the solar system for life to exist?

9 Which feature of Saturn makes it easy to identify?

10 Why is Pluto no longer considered to be a planet?

Exercise 8.2b: extension

Imagine that you are the first person to land on Mars. Write a 'space postcard' home, telling your family what it is like to be there.

Earth and Moon

The planet we know most about is, of course, the Earth. The Earth is a ball of rock surrounded by a thin layer of gas called the atmosphere. The atmosphere is made up mostly from the gases nitrogen and oxygen. About two-thirds of the Earth's surface is covered with water.

Did you know?

Ancient people thought that the Earth was a flat disc, like a plate. They thought that you would fall off the edge if you travelled too far. Aristotle lived in Greece in the fourth century B.C. He was a very clever thinker and very observant. He saw that when ships sailed away from land he could see them appearing to sink slowly out of sight instead of just getting smaller and smaller. He argued that this could only happen if the Earth was spherical. Although most people agreed with him, it was not until the first men went into space and saw the Earth from a distance that Aristotle's theory was proved right.

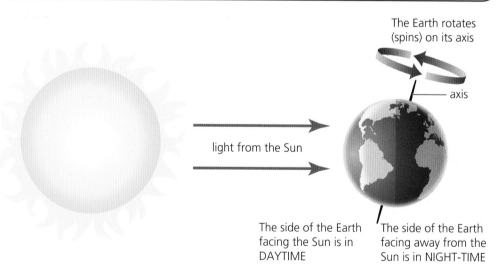

The Earth rotates (spins) on its axis

axis

light from the Sun

The side of the Earth facing the Sun is in DAYTIME

The side of the Earth facing away from the Sun is in NIGHT-TIME

As the Earth travels through space, it spins round and round on its axis. The axis is an imaginary stick passing right through the centre of the Earth. This means that each part of the Earth is sometimes facing towards the Sun and sometimes facing away from the Sun. One complete rotation (spin) takes 24 hours. We call this amount of time a day. When we are facing towards the Sun, we are in daylight and when we face away from the Sun we are in darkness—night-time.

Activity – explaining day and night

1 It is best if you can do this activity in a dark place. Use a big ball for the Earth and a lamp for the Sun.

2 Make a mark on the surface of the model Earth. Shine the light from the model Sun onto the model Earth so that the Earth has a sunny side and a dark side. Spin the model Earth on its axis. Watch what happens to the mark on the surface of the Earth. When is it in the light and when is it in the dark?

3 Use the model to help you to explain how this shows night and day and then ask your partner to explain it to you.

4 Think about the two explanations. Can you make a really good explanation by putting together the best bits from yours and your partner's?

The Earth spins round and round but the Sun stays in the same place. This means we have to look in different directions to see it. Because we cannot feel the Earth spinning, it seems to us that it is the Sun that is moving. The Sun seems to rise in the east in the morning and travel higher and higher across the sky until around midday, when it is to the south of us. It then seems to drop slowly to the west through the afternoon.

Moving shadows

When you go outside on a sunny day, you can often see your shadow on the ground. Have you ever noticed that sometimes your shadow is huge and sometimes it is tiny? Why is this?

You can show how this happens by using a torch and a shadow puppet.

Activity – daily shadows

You will need:

● a torch

● a shadow puppet or other object to make the shadow

● (a darkened room will make this activity easier).

Ask your partner to hold your shadow puppet at the edge of the table and hold the torch so that you make a shadow of your puppet on the table.

> Now move the torch up and down. What do you notice about the length of the shadow? Where do you put the torch to make the shortest shadow? Where do you put it to make the longest shadow?

When you are outside, you create a shadow by blocking the light from the Sun. As the Earth spins round, the Sun appears to move across the sky. In the early morning, it is low in the sky. We call this sunrise. Shadows are very long at sunrise.

By midday, the Sun is high in the sky, almost overhead. The shadows made at midday are short. Then the Sun begins to set and the shadows get longer again.

Sundials

Look carefully at the pictures to the right; the length of the shadow changes. What else changes?

As the Earth spins, the Sun gets higher and lower in the sky. Its position also changes. The Sun rises in the east. It is south of us at midday and then it sets in the west. As the position of the Sun changes, so the position of the shadow changes too.

Use your torch and shadow puppet again to show this. Move the torch from side to side this time. You will see the position of the shadow move.

Before clocks were invented, people used sundials to tell the time. A sundial has the time marked on the base. A triangle-shaped piece, called the gnomon, stands in the middle. When the Sun shines on the gnomon, the shadow falls on the base to show the time. As the Sun's position changes, the shadow moves around the dial.

Can you think of any disadvantages of using a sundial as a clock? Discuss this with your partner or group and then see if you could suggest what people might have done to overcome these problems.

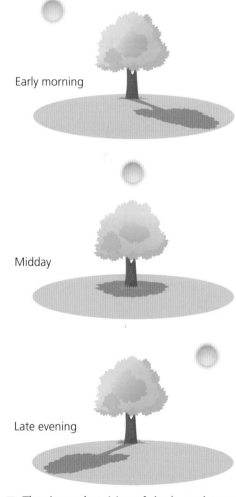

Early morning

Midday

Late evening

■ The size and position of shadows changes according to the position of the Sun during the day

■ Sundials were used to tell the time before clocks were invented

Activity – make a human sundial

You need a sunny day for this experiment.

Choose somebody to be the gnomon for the sundial.

Find a clear space where there will be sunlight all day. Ask your gnomon person to stand in the centre, or slightly to the south of the space, with his or her back to the Sun. Mark where he or she is standing.

Look at the shadow and mark the top of the shadow's head with a stick or large stone. Record the time on this marker. Make sure that no one moves these markers today.

Every hour, ask your gnomon person to stand in exactly the same place. Each time, mark where the shadow's head is and mark the time on a new stick or stone.

By the end of the day, you will have made a sundial. If you leave all the markers in place, you will be able to use it for the next few days to tell the time. Stand on the centre marker and look where your shadow falls.

■ The person's shadow indicates what time it is

Exercise 8.3a

1 How is a shadow formed?

2 Will a dark shadow be best made by a transparent, a translucent or an opaque object?

3 On a sunny day, when will your shadow be shortest?

4 Explain why shadows get longer and shorter during the day.

5 What name is given to a clock that shows the time by a shadow?

Exercise 8.3b

Fill the gaps in these sentences, using the following words:

gnomon   long   shadow   short   Sun

1 When an object blocks the light, a _____ is formed.

2 Shadows are _____ in the early morning and towards the end of the day, and _____ in the middle of the day.

3 The shadow of the _____ indicates the time on a sundial.

4 The shadow moves as the position of the _____ moves.

Seasons

The Earth takes about 365 and a quarter days to travel in its orbit all the way round the Sun. This period of time is called a year. If you look at the diagram in the 'Earth and Moon' section, you can see that the Earth is slightly tilted in relation to its orbit around the Sun, so that the North Pole is not exactly at the top and the South Pole is not exactly at the bottom. As the Earth moves around the Sun we find that the northern hemisphere (the 'top half of the Earth') is sometimes tilted towards the Sun and sometimes tilted away from the Sun. When it is tilted towards the Sun, the Sun seems higher in the sky and the light and heat are stronger. This is summer in Britain. Winter happens when the northern hemisphere is tilted away from the Sun. The Sun seems lower in the sky and the light and heat are not so strong so the weather gets colder.

Not all places have summer and winter. Places near the equator have temperatures and light that do not vary as much as they do in the UK. Can you think why?

Did you know?

The distance travelled by the Earth around the Sun is about 940 million kilometres. To travel this distance in a year we must be rushing through space at a speed of about 100 000 kilometres per hour!

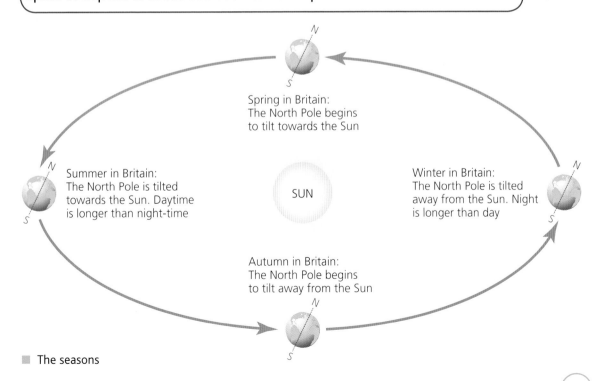

Spring in Britain:
The North Pole begins to tilt towards the Sun

Summer in Britain:
The North Pole is tilted towards the Sun. Daytime is longer than night-time

SUN

Winter in Britain:
The North Pole is tilted away from the Sun. Night is longer than day

Autumn in Britain:
The North Pole begins to tilt away from the Sun

■ The seasons

Earth's Moon

The Moon is a sphere of rock that orbits the Earth. Many scientists believe the Moon was probably formed when a lump of rock about the size of Mars crashed into the Earth about 4.5 billion years ago. It takes 28 days (a lunar month) for the Moon to orbit the Earth. An object that orbits a planet is known as a satellite. The Moon is a natural satellite of Earth. There is only one natural satellite orbiting Earth but there are thousands of man-made ones, sending pictures, global positioning data (GPS) and lots of other information to Earth.

■ The Moon

The Moon is a non-luminous object (it does not make its own light) so we can only see it because light from the Sun is reflected off its surface.

The Moon spins like the Earth so there are days and nights on the Moon too. One day lasts exactly the same time as it takes the Moon to orbit the Earth so the same bit of the Moon is always facing us. We never see the other side from Earth. The only way we can find out what the other side of the Moon looks like is to send spacecraft to take pictures.

Did you know?

The Moon is about six times smaller than the Earth so its gravitational force is six times smaller. You probably know gravity, properly known as the gravitational force, as the force that pulls everything towards the centre of the Earth. However, it is a force that acts everywhere. All the planets and their moons have their own gravity and so does the Sun. The bigger the object, the greater the gravitational force.

However, the Moon's gravity is strong enough to affect the Earth. It pulls on the water in the sea causing it to pile up more on the side of the Earth facing the Moon. As the Earth rotates and the Moon moves around the Earth, this piling up of the water happens in different places at different times. This effect causes the tides.

Activity – make an orrery

An orrery is a model that shows the positions and movements of the planets and moons in the solar system. Here are the instructions to make a simple one showing the Sun, the Earth and the Moon.

Cut a square of card that will form the base of your orrery. You could cut this from black card and draw some stars on it.

Cut a circle of paper, a little smaller than your square. Mark the centre and draw lines through this to divide the circle into four equal quarters. Then divide each quarter into three, by drawing through the centre of the circle, to make twelve equal sections. Write the names of the months in each section in order, anticlockwise around the outer edge. Stick the circle onto the centre of the black card.

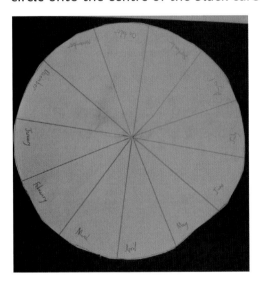

Cut a circle of card a little smaller than your paper circle. Mark the centre of the circle carefully and draw the Sun in the centre or make a Sun from coloured paper and stick it in the centre. Make a hole a little way in from the edge of this circle.

Draw a smaller circle on some card, mark the centre and then draw a little circle overlapping the edge of this circle. This little circle is the Moon, so colour it in to make it look like the Moon. Cut round the outside of this shape.

Lastly, cut out a small circle to represent the Earth. This must fit into the centre of your Moon circle. Colour it to look like the Earth and mark the centre.

To make your orrery, push a paper fastener through the centre of the Earth circle and then through the centre of the Moon circle. Push the fastener through the hole in the edge of the Sun circle and then open it up to fix these three pieces together.

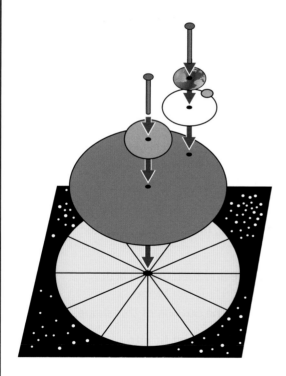

Take another paper fastener and push it through the mark in the centre of the Sun circle and then through the centre of the base card. Open it up to fix the orrery together.

Your Earth can now orbit the Sun and you can make the Moon orbit the Earth. You could draw arrows on the circles to remind yourself that they both orbit in an anti-clockwise direction.

Exercise 8.4a

1 Which two gases make up most of the Earth's atmosphere?

2 Draw a clearly labelled diagram to show how day and night are caused.

3 How many days does it take the Earth to orbit the Sun? What name is given to this period of time?

4 Explain in your own words why we have changing seasons in some places on Earth.

5 How do many scientists think that the Moon was made?

6 Explain how we can see the Moon even though it is non-luminous.

Use the words below to fill in the gaps in the following sentences. Each word may be used once, more than once or not at all.

atmosphere   axis   darkness   day   light   luminous

non-luminous   orbit   reflected   space   sphere

two-thirds   year   24 hours   28 days   365 and a quarter

1   The Earth is surrounded by a layer of gas called the _____.

2   About _____ of the Earth's surface is covered in water.

3   The Earth spins around an imaginary stick called the _____ which means that each part of the Earth is sometimes in _____ and sometimes in _____.

4   It takes _____ for the Earth to spin round once on its _____. We call this a _____.

5   The Earth takes about _____ days to orbit the Sun. We call this a _____.

6   The Moon is a _____ of rock that orbits the Earth.

7   It takes _____ for the Moon to travel once round the Earth.

8   The Moon is _____ so we can only see it when light from the Sun is _____ off its surface.

Phases of the Moon

If you look at the Moon every day, you will notice that each day it looks a bit different. It seems to get bigger and smaller. Of course the Moon does not really change in this way. The changes are caused by the fact that we can only see the part of the Moon that is reflecting the light from the Sun.

Every 28 days, as the Moon travels round the Earth, different amounts of its surface are illuminated, depending on its position relative to the Sun. Sometimes the side of the Moon facing Earth is fully lit and sometimes only part of it is lit. The different views that we see are known as the phases of the Moon.

■ The Moon seems to get bigger and smaller every 28 days as it orbits the Earth

⇨ People in science: Galileo, Newton, Einstein and the theory of gravity

Earlier in this chapter you learnt that the planets are kept in orbit round the Sun by the force of gravity. Our knowledge of how this force works has changed over the years, thanks to the work of many scientists. Scientists are still studying how gravity works and we know that there is still much more to understand. Here are three of the most famous of these scientists.

Galileo Galilei (1564–1642)

We have already met the Italian scientist, Galileo, earlier in the chapter. You will remember that he was one of the first people to use telescopes to observe the planets and moons of the solar system. His studies of the universe convinced him that the theory that the Earth was the centre of the universe and everything orbited round the Earth could not possibly be correct. He made many careful observations and tried experiments to help him to understand how the universe worked.

He knew that things fall towards the centre of the Earth, and it is said that he tried an experiment in the Italian town called Pisa. Pisa has a tower, which is famous because it leans over, making it a perfect place to try experiments involving dropping objects. Galileo's experiment was to drop several objects of different sizes and masses at the same time. Most people still expect smaller objects to fall slower than larger ones but Galileo's experiment shows that this is wrong. All the objects fall at the same rate and he realised that this must be because the force acting on them is the same. This force is what we call gravity.

■ Leaning tower of Pisa

Isaac Newton (1643–1727)

Isaac Newton was an English scientist who had an interest in many different aspects of science, including the way forces work and how this affects the movement of objects. His three 'Universal Laws of Motion' are still considered to be the basis of our understanding of the relationship between forces and motion. He was born in a village called Woolsthorpe, in Lincolnshire and you can visit his house, which is now a museum.

He studied at Cambridge University but it was when he was at home in Lincolnshire during the plague years that he came up with an explanation for Galileo's observations in Pisa and how this could explain the movement of the planets.

■ Sir Isaac Newton

There is a story that he saw an apple falling from a tree in his orchard. This helped him to realise that all objects must create an attractive force, the size of which depends on the size of the object. Newton called this force *gravitas* but we now know it as gravity.

Albert Einstein (1879–1955)

Albert Einstein was German, although he lived for most of his life in Switzerland and then the USA. Some people consider him to have been the cleverest person who ever lived because of his wide-ranging and extraordinary contributions to science. Surprisingly, he was not considered brilliant at school or in his early university studies, and he started working as a teacher. It seems that he was not very good at teaching and he moved to an office job before becoming a lecturer and then professor at various universities. He was very observant

■ Albert Einstein

and thought a lot about the things he noticed. He carried out a lot of 'thought experiments' to try to help him to understand his observations.

Einstein realised that, although Newton's laws were very good at explaining motion on Earth, they were less successful when describing the ways that all the various forces in the universe work together. He came up with a new theory that built on Newton's work. This is known as the 'Theory of General Relativity' and this and his other work earned him a Nobel Prize. Einstein's theory helped scientists to make better sense of their observations of the universe but it is still not the whole story, and so the study of gravity continues.

9 Forces

⇨ What is a force?

In Year 3 you learnt about two different types of force: magnetic forces and friction. You learnt that magnets can push or pull each other and can pull magnetic materials towards them. Magnets can also be used to stop things moving, for instance when they are used to keep something stuck to the door of your fridge. Friction can slow down moving objects or stop them moving altogether.

A force is a push, a pull or a twist applied to an object.

Forces can:

- start something moving
- slow down or speed up a moving object
- stop a moving object
- change the direction in which an object is moving
- change the shape or size of an object.

Can you think of some examples for each of these? The picture below may give you some ideas.

Forces are around us all the time. Some of them are easy to spot, for example you can see when someone is pushing down on a computer key or pulling a door open. Others, such as friction, cannot be seen so easily.

⇨ Different types of force

There are many different types of force. Here are some that you should already be able to recognise from your previous work.

Magnetic forces: magnets pull (attract) magnetic materials towards them. They can also pull and push (repel) other magnets. Can you remember how?

Gravitational force: gravity is the rather mysterious force that pulls objects towards each other. You learnt a bit about this force in the last chapter. The bigger the object, the greater the gravitational force.

Earth is a huge object, so it exerts a comparatively large gravitational force that pulls everything towards the Earth's centre. This is why, wherever we are in the world, when we drop something, it always falls downwards towards the centre of the Earth. Gravity also keeps the planets in orbit around the Sun and the Moon around the Earth.

Friction: friction is the force that is created when surfaces rub against each other. Sometimes this can be helpful and sometimes it can be a problem. Can you give examples of each? Can you say what type of surface creates the most friction? Friction causes moving objects to slow down.

Air resistance and water resistance: air resistance or drag is the frictional force created by the air as things move through it. Water resistance is the same force created in water. Objects that need to move through air or water quickly will be designed with a shape that allows the air or water to move smoothly round them. This is called streamlining.

Parachutes

Look carefully at this picture of a parachute. Discuss it with your partner or group. What do you notice about it?

When we think of parachutes we usually think of ones like this.

Use the internet to find out a bit more about the shapes of parachutes. How many different shapes can you find? Discuss the differences with your partner or groups. Are there any features that are similar in all the parachutes?

9 Forces

140

Parachutes differ in a number of ways. They may be different shapes, different sizes or made from different materials. However they all have the same job to do. They have to keep something heavy from falling too fast.

Plan an investigation to find out what type of parachute keeps a load in the air for longest. You should choose one feature to investigate (shape, size or materials). This is your independent (input) variable. Select about four different shapes, sizes or materials to test.

When you have chosen your independent variable you need to think about which variables you will need to keep the same. These are your control variables. Decide what value you will give each of these, for example, you will need to keep the load the same on each parachute. How heavy will you make it?

Your dependent (outcome) variable is what you measure to get your answer. In this case it will be the time taken for the parachute to reach the ground when dropped. You need to make sure that you have decided how you will know when to start and stop the timer and then keep this the same for all the parachutes.

It is always a good idea to repeat your tests several times (usually three). This allows you to check the reliability of your results. If all the readings for one parachute are the same or similar we say that they are reliable. If they are all very different they are unreliable. Usually we calculate the mean (average) of the three results to give us one answer for each parachute.

Draw up a table to record your results. It should look like the one shown here. The heading in the first column will depend on what question you are investigating. Your teacher will help you with this.

	Time taken for parachute to fall, in seconds			
	1st drop	2nd drop	3rd drop	Mean

When you have finished doing your experiments draw a bar chart to show the mean time taken to drop for each parachute. How can you tell which one was the best? Use your scientific knowledge to suggest why this one was better than the others.

Now discuss your experiment. How reliable were your results? Is there anything that you could have done to make your investigation better?

Did you know?

The friction between the air and a moving car slows the vehicle down. Racing cars need to go as fast as they possibly can. However a car moving too fast is in danger of lifting off the road surface into the air! Engineers need to design the cars so that the air moves smoothly past the car to reduce friction (drag) and passes over the top to keep the car on the road. We call this streamlining. They test the airflow in a special structure called a wind tunnel. Air is blown past the car very fast and smoke is added to show how the air is moving.

More about friction

The 'rule book' for all road users in the UK is called the Highway Code. Everybody who learns to drive has to learn these rules and show that they can apply them when driving. Other people, such as cyclists and horse riders should know them as well so that they can use the roads safely.

An understanding of friction is very important to drivers. The brakes on a car work by pressing blocks of metal, called brake pads, against a special surface on the wheels of the car. The two surfaces rub together and the friction that this causes makes the car slow down or stop.

There are various things that might affect how quickly the car stops and this could be really critical if the driver was trying to do an emergency stop. If the road is wet or icy, there will be less friction between the tyres and the road surface and the brakes may also become more slippery. What difference will this make to how quickly the car stops? You should be able to explain this from your knowledge of friction forces.

The distance a car travels between the time when a driver sees a danger and the time when the car stops completely is called the stopping distance. On wet or icy roads the stopping distance will be greater because of the reduced friction. What can drivers do to make sure that they are still safe and are able to stop quickly enough if they need to?

Another factor that affects stopping distances is the speed of the car. The Highway Code tells us what the stopping distances would be for a car travelling at different speeds on a dry road.

Typical Stopping Distances

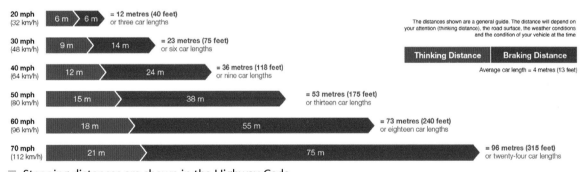

Stopping distances are shown in the Highway Code

Look carefully at the diagram. It shows us, for example, that a car travelling at 30 miles per hour (48 km per hour) would travel about 23 metres before it stopped in an emergency. The Highway Code shows us that this is made up of 9 metres thinking time and a braking distance of 14 metres. The thinking time is the length of time it takes the driver to react to the danger and apply the brakes.

The braking distance is how far the car travels after the brakes are applied. If this car, travelling at 30 miles per hour, was driving on an icy road, which part of the stopping distance would change, the thinking distance or the braking distance?

What happens to the stopping distance if the speed of the car increases to 50 miles per hour? Can you use this to suggest why the speed limit in towns and villages is usually 30 miles per hour?

So, what can we do to keep drivers and other road users safe in wet or icy conditions? The first thing is to try to increase the friction between the tyres and the road. You may have seen lorries spraying salt and grit onto icy roads. How does this help increase the friction? Drivers also have to keep their tyres in good condition and, in really icy or snowy conditions, some drivers may use special winter tyres or strap chains to the tyres to give them more grip. However, the stopping distance information in the Highway Code shows us that it is also important to remember not to travel too fast in difficult conditions.

■ Snow chains increase the grip of the when in snowy weather

Exercise 9.1

Use the words below to complete the following sentences. Each word may be used once, more than once or not at all.

brakes   decrease   direction   force   friction   reduces

gravitational   increase   pull   push   speed   stop   twist

1   A force is a ___push___ , a ___pull___ or a ___twist___ that is applied to an object.

2   Forces can change the ___direction___ or ___speed___ of a moving object.

3   ___brakes___ is a force that slows down moving objects.

4   Magnets can ___force___ magnetic materials towards them.

5   The ___gravitione___ force pulls everything towards the centre of the Earth.

6   Icy roads ___decrease___ the stopping distance of a car because the ice ___increase___ the friction between the tyres and the road.

⇨ More forces

There are many other forces that we experience every day. Here are some of them.

Elastic forces

We often use the word elastic as a noun to mean the stretchy tape that is often used in clothing. However, scientists use the word elastic as an adjective to describe materials, like the rubber in a rubber band and the foam in a sponge. They return to their original shape and size after being stretched or squeezed.

A force meter (or newton meter) is used to measure forces. It has a spring inside it. Springs are also elastic; they can be stretched or compressed and then return to their original shape when the force is removed. Force meters should be treated carefully because they can be damaged if too much force is applied. Let's find out a bit more about the forces made by things that can be stretched and squeezed.

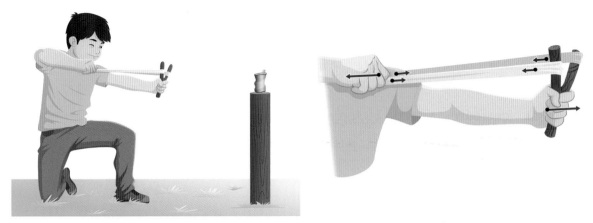

■ When we stretch an elastic material, we can feel it pulling back

Activity – stretching and squeezing

You will need:

- a rubber band
- a sponge.

Loop the rubber band around one finger of each hand. Keep your hands well away from your eyes in case the rubber band snaps or comes off your fingers.

Gradually pull your hands away from each other so that you stretch the band. Think about what you can feel. Then move your hands together and take off the rubber band.

You are using your muscles to pull on the band to stretch it. What is the rubber band doing? You can probably feel it pulling your hands towards each other. It is reacting to the force that you are applying and making its own force in the opposite direction.

Now take the sponge and push down on it gently or squeeze it so that you are pushing it inwards in several directions. Think about what you can feel. What will happen when you stop pushing?

The sponge is also reacting to your pushing force. It pushes back to return to its original shape. See if you can spot other examples of pushes and pulls made by things that have been stretched or squeezed.

Some materials will behave like elastic for a while but then break if they are pulled too much. When this happens we say that they have passed their elastic limit. We can do experiments to find the elastic limit of materials. You will learn more about this later but here is a simple experiment to start with.

Activity – find the elastic limit

You will need:

- a retort stand and clamp
- a spring
- a pile of newspapers or a piece of foam
- safety glasses or goggles
- a metre ruler
- a 100 g mass hanger
- 100 g masses.

Set up your retort stand with the clamp near the top.

Loop your spring over the bar of the clamp. Place the retort stand near the edge of the table so that the spring is hanging over the edge.

Put the foam or pile of newspapers on the floor under the spring. You will be hanging a large amount of weight on the spring and this padding will help protect the floor if the apparatus all crashes down. Remember to keep your feet out of the way!

Put on your safety glasses or goggles and make sure that you keep them on throughout the experiment. If your spring slips or breaks it could flick up into your face and damage your eyes.

Using the metre ruler, measure the length of the spring. You will need to decide exactly where on the spring you will measure and then make sure that you measure to the same place at each stage in the experiment. Record your measurements carefully in a table.

Hang the 100 g mass hanger carefully on the spring. Wait for it to stop moving up and down, and then measure the length of the spring again and record the new length.

Take the mass hanger off and check that the spring returns to its original length.

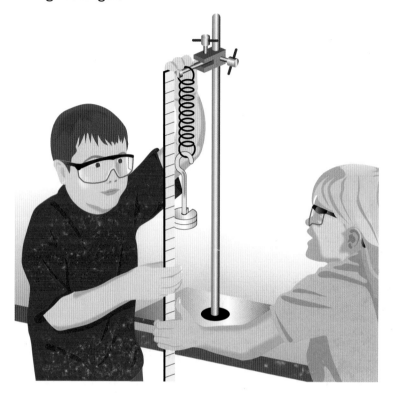

Add another 100 g mass to the mass hanger, hang it on the spring and measure and record the new length. Check that the spring returns to its original length when the hanger is removed.

Repeat this process, adding one more 100 g mass each time and remembering to check whether the spring returns to its original length. One of you should hold the retort stand firmly to stop it toppling over.

Eventually you will find that when you remove the mass, the spring does not go back to its original length but remains a bit stretched. Make sure that you make a note of when this happens. What mass on the spring caused it to pass its elastic limit?

It is quite fun to go on adding more mass to see what happens. Do it carefully, making sure that you keep your face and feet away from the spring as it may suddenly fail and spring back into your face, dropping the masses onto the floor. Measure the spring accurately each time and record the result. You do not need to take the masses off each time now since you know that the elastic limit has been passed.

Plot a line graph with 'Mass added, in g' on the horizontal axis and 'Length of spring, in cm' on the vertical axis. What do you notice about the shape of the graph? If you have done your measurements accurately, you should find that it starts off as a straight line. What happens to the line when the elastic limit is reached?

When you have tried this with a spring, you could try the same method to test other elastic materials, such as a rubber band, some elastic tape or maybe something like a jelly sweet in the shape of a loop, or some bubblegum.

Reaction (support) forces

When you stretch or squeeze elastic objects you can feel the forces exerted in reaction to the forces made on them. It is easy to feel these forces in something such as a rubber band or a spring but there are other reaction forces we take for granted and might not even know are there.

Do you remember the story of *Goldilocks and the Three Bears*? In the story, Goldilocks tries out the chairs in the bears' house, sitting on each chair in turn. The first two chairs support her, but when she sits on Baby Bear's chair it breaks. Let's think about what happens.

The Earth's gravitational force pulls everything downwards. This force keeps Goldilocks on the ground and stops her from floating away into space. It also makes her push down on any surface that is supporting her. This downward force, like other forces, is measured in newtons (N).

Goldilocks' has a downward force of 350 N. Daddy Bear, a full grown grizzly bear, has a downward force of 2500 N and Mummy Bear has a downward force of 1500 N. Their chairs are very strong. Baby Bear is much smaller, with a downward force of 300 N.

When Goldilocks sits on Daddy Bear's chair, her downward force pushes on it but she does not fall through. The chair supports her. This means that the chair makes a reaction force, pushing upwards to balance Goldilocks' downward force. Baby Bear's chair is not able to make such a big upward force. Goldilocks' downward force is too much for it, so it breaks.

Upward force is exerted every time an object is placed on a surface. It is sometimes called a reaction force but is also known as the support force. Some surfaces can exert large support forces; some can only exert small ones. For example, even the heaviest person can walk on a pavement without falling through, but soft mud is unable to support our weight very well, and no one can walk on water.

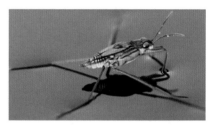

■ Water cannot support a person but some insects easily walk on the surface

Upthrust

Water exerts a support force that has a special name; it is called upthrust. Think about a boat floating on a lake. The gravitational force pulls the boat downwards, but this is matched by the upthrust of the water pushing upwards on the bottom of the boat. Of course, if the boat gets too heavy, the water can no longer support its weight and the boat will sink.

■ The gravitational force of the canoe is shown by the red arrow, and the upthrust of the water is shown by the blue arrow

Even if an object sinks in water, the upthrust works with the water resistance to slow its downward movement. If we hold heavy objects in water they seem much lighter thanks to the upthrust.

Activity – forces in water

You will need:

- a variety of objects (these will get wet so choose ones that will not be damaged by water)
- string or a netting bag to put the objects in
- a force meter
- a bowl of water.

Draw a table, like the one below, to record your results.

Object	Downward force, in N	Reading in water, in N	Does the object float or sink?

Start by hanging one of the objects from the force meter, using the string or netting bag to attach it to the hook on the force meter. You are measuring the downward force of the object in newtons (N). Record this carefully in the table. Now lower each object carefully onto the water. Do not let the object touch the bottom of the bowl. Look carefully at the reading on the force meter and record it carefully in the table.

Repeat this with the other objects, recording your results each time.

Now put all the objects into the water and record whether they float or sink.

Look at your results. What do you notice? Is there any pattern? Can you explain what is happening? What happened to the reading when something that floated was put into the water?

When an object is put into water, the upthrust works against the gravitational (downward) force. If the object floats, the two forces exactly balance one another. The gravitational force is cancelled out by the upthrust, so the reading on the force meter in the experiment will be zero. If the object is too heavy, the upthrust cannot cancel the gravitational force entirely so the object sinks. The reading on the force meter will not be zero but it will always be less than the reading taken when the object is not in the water.

Did you know?

After people have had an operation, an injury or a serious illness they often need to do exercises to help them to build up their strength. Sometimes, if they are especially weak, they may start with exercising in a swimming pool so that the upthrust of the water reduces the strain on their muscles.

Exercise 9.2a

1 What do scientists mean when they use the word 'elastic'?

2 Explain clearly why it is important not to pull the spring in a force meter too strongly. Use the term 'elastic limit' in your answer, making it clear what this means.

3 (a) Explain why a book that is placed on a table does not fall through it.

   (b) Explain in terms of support forces, why it is dangerous to walk on ice on a pond.

4 What name is given to the support force made by water?

5 Look at the canoe picture again. Explain, in terms of forces, how it floats.

6 The human body floats naturally. When people first started to spend more than a few minutes under the water, they wore diving suits with heavy boots. Explain how this helped them.

■ Early divers needed boots weighted with lead

⇨ Levers, pulleys and gears

We often need to use forces to get something done. We lift bags of shopping. We pull doors to close them. We twist a tap to turn it on. We push the pedals on a bicycle to make the wheels turn. If the force is not too great, we can do this easily. When big forces are needed we sometimes need to use something to help us achieve the task.

Levers

Levers are sometimes described as force magnifiers. What do you think this might mean?

Many everyday tools are levers. A lever is a tool that allows you to apply a force further away from where it is needed. By doing this you increase the effect of the force.

For example, it is often really hard to unscrew a nut or bolt with your fingers so we might choose to use a spanner to make the job easier. The spanner is long. It grips the nut or bolt at one end and you push the other end. The extra length of the handle of the spanner makes the job easier.

Sometimes levers are used to lift heavy objects like rocks. A long plank of wood is pushed under the rock and then something like a log is put under the plank. This forms a fixed point, called the fulcrum or pivot, and the plank can move up and down on it like a see-saw. At one end the rock is pushing down (the load). The worker then pushes down on the other end (the effort), as far away from the fulcrum as possible and can lift the

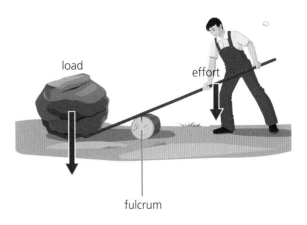

load

effort

fulcrum

heavy rock quite easily. The longer the lever is, the easier the task
will be for the worker.

Activity – other levers

Can you think of any other jobs that are made easier with tools used as
levers? Discuss with your partner or group and see how many you can
think of. You could draw pictures to make a poster of your ideas.

Pulleys

Have you ever watched a crane at work? A crane is
a machine that uses a mechanism called a pulley. A
pulley is a kind of wheel that is fixed in place. A rope is
passed over the wheel. Because the wheel turns, the
rope can move easily and smoothly when it is pulled.

A pulley allows the direction of a force to be
changed. A simple pulley might be used to lift a
heavy bucket on a building site. The bucket makes a
downward force due to gravity and normally to lift it
you would need to make an upward force. By using a
pulley you can lift the bucket with a downward force,
which is much easier.

■ Can you spot the pulleys on this crane?

If the load is really heavy, the lifting can be made
easier by using two pulleys, a fixed one at the top and
another one at the bottom, which moves with the
load. This mechanism is called a block and tackle. If
you have ever been in a sailing boat, you may have
seen a block and tackle being used to pull the sail up
the mast.

■ Pulleys on sailing boats make it easier
to raise the sail

Gears

A gear is another mechanism that makes it possible to change the
amount of effort it takes to make something happen. Gears are
found in all sorts of machines, clocks and even toys. A gear is a
wheel with little teeth all round its rim. One gear by itself is not
much use but by putting two or more together we can make big
differences to the effect of a force. To understand how gears work
we will look at a bicycle.

■ A bicycle may have many different sized gears on the back wheel

The gears on a bicycle are found in two places; the back wheel and the pedals. The gears are joined by the chain.

This diagram shows how the bicycle works. When the pedals are turned, the front gear turns and the teeth push and pull on the chain, making it move. As the chain moves it pushes and pulls on the teeth of the back gear and makes it go round. In this first diagram, the two gears are the same size, every time the pedals turn the front gear once, the back gear will turn once and so will the back wheel.

In the second diagram, the two gears are very different sizes. The small gear has fewer teeth than the large one. When the pedals turn the larger gear round once, it will make the smaller gear go round several times. This would be very hard work if you were going uphill but on the flat or downhill you could go really fast.

In the third diagram, the gear on the pedals is smaller than the one on the back wheel. The cyclist would have to turn the pedals several times to make the back gear and the back wheel turn once. This would make it much easier to pedal uphill. The cyclist would have to turn the pedals quickly but it would not be such hard work.

If you have a bicycle with gears, have a good look at them. Many people have gears but do not know how to use them to make cycling easier. Now you know how they work you should be able to make the best use of gears on your bike in the future.

Did you know?

Racing bikes have several gears on the pedals as well as on the back wheel so they can have lots of different combinations to choose from to help riders to go as fast as possible.

Exercise 9.3

1 What term is sometimes used to describe a lever?

2 What is the name of the fixed point around which a lever moves?

3 Which would make a lifting job easier, a long lever or a short one?

4 Explain how a pulley makes it easier to lift a heavy load.

5 What name is sometimes given to a combination of two or pulleys?

6 What is a gear?

7 Give two places where you might find gears being used.

⇨ Force diagrams

When scientists record their findings or observations, they often do so in the form of a diagram. We can use diagrams as a way of showing, quickly and simply, how a force is acting on an object. This is done using arrows. You may have noticed the arrows on the diagrams of the different forces we have been looking at in this chapter. The direction of the arrow shows the direction in which the force is acting and the length of the arrow can be used to represent the strength of the force. The tail of the arrow should always be at the place where the force is acting.

Here is an example.

The book rests on the table and the table supports it. The downward or gravitational force, is balanced by the upward support force made by the table. The arrows are therefore the same length.

support force

gravitational force

Notice that the arrow showing the weight (gravitational force) starts right in the middle of the book. This spot is called the centre of gravity and is where we consider the downward force of the object to act.

Here is another example of a forces diagram.

In this picture, one man is stronger than the other. His pulling force is greater so the arrow showing his force is longer. The forces are therefore unbalanced and the stronger man will pull the weaker one across the ground.

Activity – draw force arrows

You will need:

● pictures showing forces acting on objects (you could draw these, cut them from newspapers or magazines or maybe your teacher will give you some)

● a ruler

● a pencil or coloured pencils.

Look carefully at each picture and identify the forces acting on the object. Think about whether the forces are balanced (the forces are equal) or unbalanced (the forces are unequal). If they are unbalanced think about which force is bigger.

Use your ruler and pencils to draw arrows on the pictures to show the forces acting. Remember to make the length of the arrow show the relative strength of the forces.

Exercise 9.4

Use the words below to complete the following sentences. Each word may be used once, more than once or not at all.

arrows   direction   equal   gravity   length   longer   mass

point   shorter   tail   unbalanced   upthrust

1 Forces can be shown on diagrams using _____.

2 The direction in which the force is acting is shown by the _____ of the arrow.

3 The relative strength of the forces is shown by the _____ of the arrow.

4 If the forces are _____ , the arrows will be the same length.

5 If the forces are unbalanced, the stronger force is shown by a _____ arrow.

6 The _____ of the arrow should be at the point where the force is acting.

7 The gravitational force is considered to act at the centre of _____.

Glossary

Absorbent   Soaks up liquid easily.

Adaptation   Feature of the body of an animal or plant that improves its ability to survive in its habitat; the process of change in features of an animal or plant to improve its ability to survive in its habitat.

Adolescence   Time when changes occur in the body as a child becomes an adult (also known as 'puberty').

Air resistance   Slowing effect caused by friction between a moving object and air.

Algae   Non-flowering plants found in water or damp places, usually single-celled.

Aluminium   Light metal used to make aeroplanes and kitchen foil.

Amniotic sac   Protective bag of watery liquid in which a fetus develops.

Amphibians   A group of vertebrate animals with moist skins that lay eggs in water, e.g. frog.

Asexual reproduction   Reproduction involving just one parent, e.g. taking cuttings from a plant.

Astronomer   Scientist who studies the stars and other objects in the universe.

Atmosphere   The layer of gases that surrounds the Earth or another planet.

Axis   Imaginary line passing through the centre of a rotating object, about which the object turns.

Bio-fuel   Fuel made from plant material.

Biodiversity   The range of different types of plant and animal found in a particular habitat.

Breasts   Parts of a woman's body that make millk to feed a baby.

Brittle   Breaks easily into pieces.

Bulb   A structure of some plants, e.g. bluebell, that is found underground and which stores energy.

Camouflage   Features of the body (shape and/or colour) of an organism that allow it to blend into its background.

Canopy   The upper layer of a woodland formed by the branches of trees.

Cells   The building blocks of living organisms.

Centre of gravity   Point at which the downward gravitational force of an object is considered to act.

Cervix   The opening to the uterus.

Chemical reaction   Non-reversible change that occurs when a substance is heated or mixed with another substance, resulting in the formation of new substances.

Chromatogram   The piece of paper showing the result of separating a mixture by chromatography.

Chromatography   Method used to separate mixtures of dissolved substances e.g. coloured dyes.

Combustion   Burning.

Community   The group of animals and plants that live in a particular habitat.

Component   Part of a mixture.

Condense   Change from a vapour to a liquid when cooled.

Constellation   Group of stars forming a recognisible pattern in the sky.

Contract   To get smaller.

Control variable   Variable that must be kept the same during a fair test.

Copper   Brownish metal used to make water pipes and electrical wiring.

Decant   Method of separating a mixture of a liquid and a heavy, insoluble solid by pouring the liquid into a separate container leaving the solid behind.

Dehydrated   Lacking in sufficient water.

Dependent (outcome) variable   The variable in an investigation that is measured to obtain the results.

Diagram   Representative drawing e.g. showing how apparatus is set up.

Dissolve   To mix completely with a liquid and seem to disappear.

Distillation   Method of separating a mixure by evaporation and condensation of one or more liquids.

Distilled water   Water that has been made pure by distillation.

Dye   Substance used to make something coloured.

Egg   Female reproductive cell (ovum).

Effort   Force applied to a moving object such as a lever or pulley.

Elastic   Able to return to its original shape and size after stretching, compressing or twisting.

Electrical conductor   Material that allows electricity to travel through it easily.

Electrical insulator   Material that does not allow electricity to travel through it easily.

Evaporate   Change from a liquid to a vapour when heated.

Expand   To get larger.

Extinct   There are no more living examples of the species.

Fallopian tube   The tube leading from an ovary to the uterus (also known as 'oviduct').

Fertilisation   The joining together of an egg cell and a sperm cell to form a single cell that can develop into a baby.

Fetus   Developing baby.

Filter   To remove an insoluble solid from a suspension by pouring it through filter paper; device containing filter paper that is used for filtration.

Filtrate   The liquid that has passed through a filter.

Fire blanket   Non-flammable sheet of fabric used to put out a fire by removing the supply of air.

Fire extinguisher   Device used to extinguish (put out) a fire.

Fire triangle   The three components (fuel, air supply and heat) that are necessary for combustion to occur.

Fledge   (of young birds) Leave the nest.

Foam   Material containing trapped bubbles of air or other gas.

Force   Push, pull or twisting action causing an object to move or change shape.

Fossil fuel   Fuel formed deep underground millions of years ago from decayed plants and animals; coal, oil and natural gas are examples of fossil fuels.

Freeze   Change from a liquid to a solid when cooled.

Friction   Force caused by two objects rubbing together.

Frog spawn   The eggs of a frog.

Fuel   Material that can be burnt to provide thermal energy.

Fulcrum   Fixed place about which a lever acts.

Galaxy   Large group of stars e.g. Milky Way.

Galvanisation   Protection of iron objects by coating the surface with another metal, ususally zinc.

Gamete   A sex cell, either an ovum (egg) or a sperm.

Gear   A toothed wheel used to change the relationship between effort and movement, e.g. on a bicycle wheel.

Geology   Study of rocks.

Gestation Period   Length of time a fetus spends developing in the uterus.

Gills   Structures that allow water animals to extract oxygen from the water.

Gnomon   The part of a sundial that casts a shadow used to tell the time.

Gravitational force   The force acting between two objects and pulling them together, gravitational force (sometimes just called gravity) keeps planets and moons in orbit and pulls dropped objects towards the centre of the Earth.

Habitat   The area where an animal or plant lives and which provides food, water and a place to breed.

Hemisphere   Half of a sphere.

Herbivorous   Plant-eating.

Hibernate   To slow body processes to a complete minimum and sleep throughout the winter.

Imago   The flying adult form of an insect such as a butterfly.

Independent (input) variable   The input variable in an investigation, the values of which are selected by the researcher.

Insect   A group of invertebrate animals that have three body parts and six legs, e.g. butterfly.

Insoluble   Not able to dissolve.

Lever   Rigid bar resting on a fulcrum used to make it easier to lift heavy loads.

Life cycle   The different stages of the life of an organism.

Limewater   Clear liquid used to show the presence of carbon dioxide (turns cloudy).

Load   Heavy object or force acting on an object.

Lubricant   Material used on the surface of moving objects to reduce friction e.g. oil or grease.

Luminous   Producing and giving out light.

Magnetic   Attracted to a magnet.

Mammal   Group of vertebrate animals which feed their young on milk and have hairs on their bodies.

Man-made   Made by humans by mixing or changing materials.

Man-made material   Material that is created by changing natural materials e.g. by heating.

Mass   A measure of how heavy something is, measured in g and kg.

Material   What an object is made from.

Melt   Change from a solid to a liquid when heated.

Menstrual cycle   Monthly cycle during which the lining of the uterus is shed and renewed.

Metamorphosis   A complete change of body form during a life cycle, e.g. tadpole to frog.

Migrate   To move from one place to another during the year, usually for food or to find suitable breeding grounds.

Migration   A movement from place to place, usually in search of fresh food, or to a breeding area.

Milky Way   The galaxy that contains the Solar System.

Minerals   Substances absorbed in small quantities from the soil by the roots of plants to provide materials for healthy growth.

Mixture   Two or more materials mixed together but not chemically combined.

Moon   A rocky sphere orbiting a planet.

Natural material   Material that is found in a usable state in nature.

Nerves   Special cells which carry messages around the body.

Nitrogen   Gas that makes up about 80 per cent of the air.

Nocturnal   Active at night.

Non-luminous   Does not produce and emit light.

Non-magnetic   Not attracted to a magnet.

Non-reversible   Not able to be reversed (turned back).

Opaque   Light cannot pass through.

Orbit   To travel around; the path taken by an orbiting object.

Organism   A living thing.

Ovaries   The parts of a woman's body that make the eggs.

Oviduct   The tube leading from an ovary to the uterus (also known as 'fallopian tube').

Ovum   Egg, female sex cell.

Oxygen   Gas, necessary for combustion and respiration, which forms about 20 per cent of the air.

Penis   Part of a man's body through which sperm and urine leave the body.

Period   The part of the menstrual cycle when the lining of the uterus is shed and bleeding occurs.

Phases of the Moon   Apparent different shapes of the Moon seen as it orbits the Earth, caused by different amounts of the lit side of the Moon being visible from Earth.

Photosynthesis   The process used by plants to make their own food from water and carbon dioxide, using light energy from the Sun.

Pioneer   A person who takes the lead into a new area or is the first to try a new activity or piece of equipment.

Pipette   A piece of apparatus, this is a tube with a bulb at the end, used to suck up and transfer small quantities of liquid.

Pivot   The fixed point at which a lever rotates.

Placenta   Structure joining the fetus to the wall of the uterus through which oxygen and nutrients pass from the mother's blood to the fetus's blood and waste substances pass in the opposite direction.

Planet   Object orbiting the Sun or another star, e.g. Mercury, Earth.

Plankton   Tiny plants and animals that live in water.

Plastic   Able to be moulded into different shapes; a man-made material, usually formed from oil, that is easily shaped to make many objects.

Pollination   Transfer of pollen from the anther of one flower to the stigma of another flower of the same species.

Polluted   Containing harmful substances.

Predator   An animal that hunts and kills other animals to eat.

Prey   An animal that is hunted and eaten by a predator.

Product   New substance formed during a chemical reaction.

Property   The way a material behaves under certain conditions.

Puberty   Time when changes occur in the body as a child becomes an adult. Also known as 'adolescence'.

Pulley   Wheel around which a rope or cord can pass to allow a load to be lifted by a downward force.

Pupa   The part of the life cycle of an insect when metamorphosis takes place.

Pure   Not mixed with anything else.

Purified   Made pure.

React   To change into one or more different substances when heated or mixed with another substance.

Reaction force   Force that is exerted by an object in resistance to being pushed, pulled or twisted.

Recycle   To collect and refresh a resource, e.g. metals or paper, so that it can be used again.

Reliability   A measure of how good a set of data is. If repeated readings are close to the same value, the data set is reliable.

Reproduction   A life process that creates new individuals, usually requiring two parents.

Residue   The insoluble material left in a filter after the liquid has passed through.

Reversible   Able to be reversed (changed back).

Rigid   Unable to bend.

Rotate   To spin around an axis.

Rotation   Spinning.

Rover   A vehicle that has been delivered to another planet by a spacecraft and can move around, controlled by remote control from Earth, to explore and sample the surface of the planet.

Rust   Brown crumbly substance (hydrated iron oxide) that forms on the surface of iron in the presence of oxygen and water.

Rusting   The reaction between iron metal and oxygen in the presence of water to form rust.

Satellite An object, either natural or man-made, that orbits a planet.

Saturated solution Solution containing so much dissolved solvent that no more can dissolve.

Scavenger An animal that eats dead plant or animal material.

Sediment Insoluble material that sinks to the bottom of a container or body of water.

Seed Structure that forms from a fertilised ovum in a plant containing an embryo plant.

Seed dispersal Spreading seeds over a wide area.

Sexual reproduction Reproduction involving two gametes, one from each of two parents.

Shadow Area where light has been blocked out by an opaque object.

Sieve Device containing small holes, used to separate mixtures of different sized pieces.

Soluble Able to dissolve.

Solute Soluble solid that dissolves to make a solution.

Solution Mixture formed from a solvent and a dissolved solute.

Solvent Liquid part of a solution, usually but not always water.

Spatula A spoon-like scoop used in science to transfer solid material from one container to another.

Species A type of organism; the members of a species can breed successfully with each other.

Sperm Male sex cell.

Star Massive burning ball of gas.

Steel Metal mixture (alloy) formed from iron and carbon, sometimes including other metals, e.g. chromium.

Streamlining Shaping something so that air or water passes smoothly over the surface causing the minimum of friction.

Sun Our nearest star.

Sundial Device used to tell the time from a shadow which moves across a dial as the Sun moves through the sky.

Support force Upward force created by an object when a load is placed on it and which supports the load.

Suspension Mixture of small pieces of an insoluble solid floating in a liquid.

Sustainable Able to be continually produced without harm to the environment.

Tadpole The young of a frog.

Telescope Device used to make distant objects, e.g. stars, seem nearer so that they can be studied more easily.

Territorial   Defending a territory.

Territory   An area of habitat that is defended by a particular animal.

Testes   Parts of a man's body that make sperm cells.

Tide   The daily rising and falling of the sea caused by the gravitational pull of the Moon.

Translucent   Light can pass through but you cannot see through clearly.

Transparent   Light can pass through and you can see through clearly.

Umbilical cord   Structure which joins a fetus to the placenta.

Universe   All the matter and energy that exists.

Upthrust   Upward force created by water when an object is placed on the surface; if the upthrust can equal the weight of the object, the object will float.

Uterus   Structure in a woman's body where a fetus develops (also known as the womb).

Vagina   The part of the female reproductive system between the uterus and the outside, through which sperm enter for reproduction and a baby leaves the uterus when it is born.

Vapour   Gas formed when a liquid evaporates.

Variable   Factor that can change or can be changed in an experiment.

Volume   The amount of space taken up by an object or contained within something, e.g. a beaker, measured in cm^3 or litres.

Water resistance   Slowing effect caused by friction between a moving object and water.

Weight   The downward force created by the mass of an object and the gravitational pull towards the centre of the Earth, measured in N.

Womb   Structure in a woman's body where a fetus develops (also known as the uterus).

Word equation   Simple way of showing the reactants and products in a chemical reaction.

Year   Time taken for a planet to orbit the Sun; the Earth's year lasts for $365\frac{1}{4}$ days.

Zinc   Metal often used in galvanising to protect iron from rusting.

Zoology   Study of animals.

Index